Physics and Metaphysics of Music and Essays on the Philosophy of Mathematics

BY LAZARE SAMINSKY

BOOKS

Music of Our Day
Music of the Ghetto and the Bible
Living Music of the Americas
The New Art of Conducting

MAJOR MUSICAL WORKS FOR THE STAGE

The Daughter of Jephtha, cantata-pantomime, three scenes
First performances in Rome and New York
Published by Salabert, Paris
The Defeat of Caesar Julian, opera, three acts (in preparation)
The Vision of Ariel, opera-ballet, one act, stage première in Chicago; concert performances in Paris, New York, etc. (Mercury)
Rachel, a biblical ballet
Performances in Petrograd, Paris, New York, etc.
Published by Salabert, Paris; and Mercury, New York
The Plague's Galliard, opera-ballet, one act
First performances in New York, Paris and Vienna
Published by Salabert, Paris; and Mercury, New York

MAJOR MUSICAL WORKS FOR THE ORCHESTRA

Second Symphony (Salabert)
Premières in Amsterdam, New York, Petrograd
Third Symphony
Performances in Paris, New York, Moscow, Vienna, etc.
Published by Universal Edition in Vienna
Fourth Symphony (Universal Edition)
First performance in Berlin
Fifth Symphony (in preparation)
Ausonia, Italian Pages (Salabert)
Premières in Florence, Zurich, Toronto, Buenos Aires, etc.
Three Shadows (I. W. Chester, London)
Performances in New York, Zurich, Chicago, etc.

MAJOR MUSICAL WORKS FOR THE VOICES AND ORCHESTRA

Litanies of Women (Salabert)
Performances in Boston, New York, Paris, Berlin, Vienna, Milan, Venice, Buenos Aires, etc.
Songs of the Russian Orient (Universal Edition)
Performances in London, Berlin, Paris, Rome, Zurich, Milan, Naples, Montreal, etc.
Eon Hours (Mercury, New York)
Poems from the Edda, Charles d'Orléans, etc.
For vocal and instrumental ensemble
Première in New York

Physics and Metaphysics of Music and Essays on the Philosophy of Mathematics

BY

LAZARE SAMINSKY

A GREEN PHILOSOPHER'S PERIPETEIA
PHYSICS AND METAPHYSICS OF MUSIC
THE ROOTS OF ARITHMETIC
CRITIQUE OF NEW GEOMETRICAL ABSTRACTIONS
THE PHILOSOPHICAL VALUE OF SCIENCE

Springer-Science+Business Media, B.V.

1957

Originally published by Martinus Nijhoff, The Hague, Netherlands in 1957
Softcover reprint of the hardcover 1st edition 1957

ISBN 978-94-017-4606-9 *ISBN 978-94-017-4822-3 (eBook)*
DOI 10.1007/978-94-017-4822-3

To the memory of

DR. MORRIS RAPHAEL COHEN

Dedication

During my summer at the MacDowell Colony in Peterborough, New Hampshire – over three decades ago – among the present and future "gods" of American literature headed by the eminent American poet Edwin Arlington Robinson, I found a submerged figure: a man slight and taciturn, his eyes reflecting determined thought, but also humor of shy charm.

That was Dr. Morris Raphael Cohen, then professor of Mathematics and Philosophy at the City College of New York, working at the Colony on his famed *"Reason and Nature"*.

I met Dr. Cohen later, a number of times; admired his philosophical works. It was touching and comforting to find that in his earliest days as philosopher he selected as guides two formidable minds of our century, Alfred N. Whitehead and Bertrand Russell. Nearly thirty years after my first meeting Dr Cohen, I delivered a memorial address on him in New York, at a splendid evening devoted to the great figures of the MacDowell Colony who had passed during the preceding years.

I quoted his students calling him "modern Socrates with an acid tongue", and his philosopher friends – "the greatest American thinker since William James". He called himself "a logician". Dr. Cohen was all of this. In his genius for integrating thought with scientific evidence; in his acute analysis of meaning, reasoning and implication; in his defense of the principle of polarity, one marvels at both the originality and the precision of his thinking.

For yet another trait, Dr. Cohen's vast thought was also described as "the spirit of intelligent dissent". A mind that loves fresh avenues of thought, intellectual space and tolerance, is what civilized humanity needs so direly.

This book is dedicated respectfully and lovingly to the memory of Dr. Morris Raphael Cohen.

Rye, New York. LAZARE SAMINSKY
Summer 1955.

Contents

PRELUDE

A Green Philosopher's Peripeteia

My philosophical interest awakened early. At the beginning of this grim century I was a wide-eyed boy in Southern Russia dreaming of becoming a composer – without knowing a thing of even the elementary technical demands of the art.

I had my fingers also in philosophy and mathematics – with just a taste of both. With all the daring of ignorance I sailed into Descartes, translating his *Meditationes de Prima Philosophia* from the Latin and French (the Hague) edition; wrote a "commentary" on Spinoza's *Ethics*; was deciphering Kant's *Prolegomena zu jeder Metaphysik*; also, revelling in celestial mechanics with the first thrill of geometry of curved spaces giving it an air of romance.

The name Einstein was to be heard of by us in Russia much later. However, I already thumbed pages of Poincaré (Henri, the mathematician) and of Helmholtz, tried to unravel those of Riemann and Lobatchĕvski, one of the earliest and greatest builders of the curved space concept.

I remember vividly the later years' extracurricular lectures on the application of calculus and on mathematical logic and also, on thermodynamics (on the theory of gases and radiation) which I managed to attend at the University of Odessa not yet being a student. Much was said of Van 't Hoff and his "osmotic pressure" – not a word on Max Planck, creator of the quantum theory that was to change the face of mathematical cosmology.

My thoughts on the philosophical meaning of music also arrived early. They were fired by Schopenhauer's *Die Welt als Wille und Vorstellung*, which I was reading feverishly as a back bencher at my Imperial Lyceum of Commerce. That, too, happened in handsome, light-minded Odessa on the Black Sea ... I heard then of Pythagoras' acoustics and of his philosophy of music. Was mightily impressed by the Leibnitz dictum: "*Musica est exercitium arithmeticae occultum nescientis se numerare animi*" (Music is a veiled excercise in arithmetic by the soul unaware of its counting) [1]. I had not quite realized, at that time, that here we have merely a statement on acoustics, not a philosophy of music.

But then, for many years my thoughts of the cosmic significance of music had gathered in my mind in only a rambling way. Committed to notes left for years unrelated and unorganized, these thoughts were, however, driven by a belief in music's being in some way a hint at the Ultimate Real.

Passionately I plunged into mathematics of curved spaces, reading among others Bertrand Russell's *Foundations of Geometry* ... My new knowledge I was using for the defense of a fiercely Kantian point of view against philosophizing mathematicians. These, I insisted, take in – in their generalizations – too much territory and invade the epistemological realm with armament both slender and improper.

That activity absorbed much of my thought. But then, a very intense composer and conductor life, now entirely professional, kept me for years from going back to my mathematical and philosophical work, as well as from formulating my own views on the inner nature of music.

A few years before my trip to the North to become a student at both, the University of Saint Petersburgh (now Leningrad) in mathematics, and of the Imperial Conservatory of Music – in composition and conducting, I began to fashion my first original work *The Philosophical Value of Science.*

Having written this piece, and doubting my competence in either mathematics or philosophy – I was nineteen – I kept the finished work for over a year in my portfolio. Then, in a fit of daring I sent it to the Moscow University magazine "Problems Psychology and Philosophy," then one of the greatest philosophical journals in the world. I still felt that my work could not be satisfactory, but I was longing for someone's advise and judgement.

After a few months of nervous waiting, a letter arrived from the editorial board, composed of some of the greatest mathe-

[1] Leibnitz in a letter to Goldbach of 1712.

maticians and philosophers of Russia. In delicate and gracious terms the august scholars said: "Oh yes, you *are* a greenhorn." ... They praised me for "thoughtful treatment of my subject" and for "correct and valuable ideas" mingled, however, as they put it, with overlong passages showing an insufficient and elementary training in either philosophy or mathematics.

Curiously enough, I did not feel defeated; I rather felt elated, and set out again to storm the gates of science.

The grand scholars shrugged also at my "dissertation" on the nature of algebra as a vision of permutations which exhausted the mathematical potential of a set of algebraic propositions. This idea of mine was, however, not entirely naive nor without a future. I was greatly elated reading, decades later, the definition of multiple algebra as "calculus of implications."

Cheerful it was, again decades after my green meditations on algebra, to come upon the name and work of Alfred North Whitehead and find this statement:

"The ideal of mathematics should be to erect a calculus to facilitate reasoning in connection with every province of thought, or of external experience, in which the succession of thoughts, or of events can be definitely ascertained and precisely stated." [1].

Nor was left unsustained my belief in application of mathematical logic and of its tools – then, in 1903, *in statu nascendi* – to my ideas of algebra's mathematical evolving of patterns and potentials of judgment. Our own day is watching new electronic computers which scan mass answers to definite problems, computers of incredible swiftness and precision. They can calculate the consequences of mathematical equations of the greatest importance to humanity in a few months while it would take several brilliant mathematicians a life time to reach and draw all the correct sequels from the same set of equations.

Here I have changed nothing in either of my two older essays published in Russia decades ago: "On the Roots of Arithmetic" and "Critique of New Geometrical Abstractions." In spite of their orthodox Kantian animadversions, these two pieces are

[1] Alfred North Whitehead: *A Treatise on Universal Algebra.*

firmly – and tellingly, I hope – related to the totality of my philosophical views. And "Physics and Metaphysics of Music" takes care of making my generalizing points clear and final.

Looking at my older essays now – the "Critique of New Geometrical Abstraction," for example – I see that mine was in one way a loosing battle: *for* the Kantian old believers, and *against* the new geometricians. "Physics and Metaphysic of Music," however, gives me a new comfort of balance. In the new mathematics and physics my philosophy of music has found a fresh reserve of energy.

In this edition I have included the *Philosophical Value of Science*, but only as an addendum. This I did for various reasons, but also, as a part of an autobiography of a mind. I have eliminated though the more elementary parts of the discussion of the nature of algebra, of the primordial instrumentality of touch, etc.

Re-reading the later essay made clear to me that, though I already knew something of mathematics of curved spaces, my fierce Kantian views were not entirely shaken later.

At that early date I insisted – in an already broader formula, that "even if mathematical concepts originated in the experience of our senses; even if mathematical truths are achieved by induction, mathematics still have their deeper source in *a priori* foresight," ... that "there is a difference between the reasoning weight of abstraction in the mathematical idea, on the one hand, and mathematical judgement – on the other."

But trying to reconcile a Kantian creed with "*my*" discovery of new geometry, I still insisted that the philosophizing mathematicians blunder away into reasoning too far from the natural habitat of their evidence.

My erstwhile scouting about in Descartes showed in a naive manner. Groping about blindly I came to the conclusion that science has to eliminate from its reasoning the very idea of energy, and that physics reduce themselves to a mere mathematical formulation of events and of the concomitants of their emergence. Which was a naive blend of neo-cartesian thought with an "anticipation" of the later Einsteinean concepts.

As I have hinted above, that green essay of mine "The Philosophical Value of Science," has peculiar points of significance – in the manner of its creative birth: one of the reasons why I am keeping it, in a somewhat condensed form, in this edition.

Many years after my toiling over the piece I became aware of the inner meaning of it, and of its position in the ensemble of my works.

That early work of mine on philosophy of science has still another interesting corrolary when we realize that new mathematics and new cosmology have returned – not entirely to Kant but quite pointedly, to some tenets of the Kantian creed. It is a comfort to underline the following passage from my oldest essay as a kind of forestalling Einstein's *profession de foi.*

"It is true that the experience of our senses could be – and, at times, has been – the inciter of mathematical ideas and perhaps also, of mathematical reasoning. It is easy to see, however, that even if its basic concepts were formed empirically, mathematics has long since abandonned that deceptive, filmy gate: the experience of senses ... *It is highly significant that a function embodying a law of nature is chosen from the ocean of mathematical propositions, not vice versa.*" [1]

And to read *now* the comparatively recent statement of Einstein repeating his old belief.

"The supreme task of the physicist is to arrive at those universal elementary laws from which the cosmos can be built up by pure deduction. There is no logical path to these laws; only intuition, resting on sympathetic understanding of experience, can reach them. In this methodological uncertainty, one might suppose that there were any number of possible systems of theoretical physics – all with an equal amount to be said for them ... But evolution has shown that at any given moment, out of all conceivable constructions, a single one has always proved itself absolutely superior to all the rest." [2]

Precisely this awareness has provided additional fuel to the thesis of my philosophy of music.

Curiously enough, the *Philosophical Value of Science* says not a word about music and its cosmic significance, the key motive of

---

[1] L. Saminsky: *Philosophical Value of Science* (in this book).

[2] Albert Einstein: *Essays in Science* (Philosophical Library, New York).

my last and most mature essay *Physics and Metaphysics of Music.*

There is, however, a bridge between them.

I have noticed that in addition to baring the root form of all my works here, "Philosophical Value of Science" shows a mode and structural pattern apparently typical of *all* of my creative work. This mode is one for either, my creative tonal thinking and philosophical. Which in itself sustains my own monistic view on the stance and substance of Being.

The *Philosophical Value of Science* has a musical counterpart in my early orchestral "triad of poems" *Vigiliae* (The Vigils). Both were born in the same period of initial creative and technical growth. Both show a parallel route in either, my musical form building and philosophical quest.

The above two works played exactly the same role in the unfolding of my twin spheres of creation; acted as the first shoots that hid, then fed a potential of growth during my best creative period.

Out of the *Philosophical Value of Science* sprang the later triad of my philosophical works; out of *Vigiliae* – the triad of my symphonies: Of the Great Rivers, the Summits, the Seas. Why it should be so, could be, perhaps, seen in the light of William James' wondrous *aperçu* as to the essence of philosophy: vision, not technique.

When in my distant youth I was getting drunk on Bertrand Russell's *Essai sur les fondements de la géometrie* – not knowing English then, I had to read his curved spaces' mathematics in its French version – I did not foresense how closely my own philosophy will get, many years later, to one segment of Russell's line of thought: on the relation between philosophy and mathematics.

Only recently and after I have designed and defended my own trivium – logic, mathematics, music, – I had read, with a deep feeling for the brotherhood of thought, the following passage in the autobiography of an outstanding American philosopher, Dr. Morris Raphael Cohen.

"Russell's demonstration that pure mathematics or logic can not be identified with either psychologic or physical events but constitutes a part of the real world of thought, seemed to me to offer a well-grounded and fruitful starting point for philosophy ... Equality is not more real than inequality, variables are not more real than constants. Accepting the ultimate validity of logic and mathematics thus made it possible for me to see that unity and plurality, similarity and difference. dependence and independence, form and matter, change and permanence, are equally real." [1]

This nearness to Russell's view quite naturally and pointedly showed in my published essays on philosophy of arithmetic and critique of the new geometrical concepts now re-published in my English translation, in this book.[2]

As the reader will see, in these essays my sternly Kantian view of early youth had softened and became open to modification. I remained adamant, however, then and for some time, on pitting the scientific-philosophical value of the new mathematical thought against its epistemological irrelevance.

Long before I went into the final shaping of *Physics and Metaphysics of Music*, Einsteinean mathematics and relativity changed the whole body of cosmological evidence that was to sustain the shaping of my own philosophical creed.

Those doors in some way open on the Whole I have been approaching ardently. And if I staggered, it was because of the ocean of contradicting theories, controversies and endless self-correcting by even the greatest in the science of cosmology.

Having no claim on original scholarship, I would not presume to go into cosmological research of my own; or even check upon the mighty. All I could do, is to confront evidence presented by great scholars – mathematicians, physicists, etc. – and hold up their discords for detached consideration. Whenever their facts or the meaning read into them disagreed, I *had to* interprete the oddities of "fact-finding" as best I could – in accordance with my own ways and views.

Also, one should not forget that philosophers in metaphysicians' class do not and can not have that awe for, or even belief in – the "outside universe" with it's "laws." And this, also in the present

[1] Morris R. Cohen: *A Dreamer's Journey* (The Free Press).

[2] *The Roots of Arithmetic* and *Critique of New Geometrical Abstractions* were first published in "Problems of Psychology and Philosophy" (Moscow).

case – of a writer fully aware of these laws being probabilities tending to a maximum, and at best – not deductive but intuitional and forecast-laden; of a writer fully aware of the functional import of the outside world as a shadow or "tangent" of the Ultimate.

Of course, all caution and delimitations aside, these are *our* metaphysics, no matter how abrtact and detached, no matter how distant from the raw "experience" and "the outside" that coarsens our higher vision.

It is my deepest conviction, however, that metaphysics able to preface itself with, sift wisely and codify the highest critique of abstractions yielded by philosophy of science, can achieve an insight into what Immanent Being might be.

Physics and Metaphysics of Music

I

We live in concentric chains of uncertainly, that is, in a dual envelope of the outer world – a fiction in itself – and of the immediate, surfacial ego full of false sensations. Our immediate self, or perceptual mind, turns on an ingrown notion that all it sees and takes in, is true. In other words, to be observed means also "to exist".

Yet the sense of "here-now" and the sense of the inherently "present" are not the same thing. One is a statement on the empirical world, the other – on the anticipated "beyond." The former, a sense-datum brought to the surfacial, the minor self; the latter, a gleam of self-transcending awareness.

The above uncertainties do whirl "around" [1] an instantaneous and reasonably "sure" awareness of the *given* which has never been better formulated than in Descartes' "*cogito ergo sum*" – "I think: that means, I am".

"Present" as such is, of course, the sharp point, or the perch, so to speak, of "*cogito ergo sum.*" Bergson's line on "reaching the fundamental self by deep introspection," in the finale of "*Données immédiates de la conscience,*" is really a restatement of Descartes' axiom.

Just as precise and perhaps even a shade more Cartesian are Eddington's variations on a Bergsonian theme: "When I close my eyes and retreat into my inner mind, I feel myself enduring, I do not feel myself extensive" [2], and: "Mind is the first and most direct thing in our experience; all else is remote inference." [3]

[1] Using the ambigous symbol "around" or, later, "wrapped about", "calm flow" (about the life of our I) etc., we must not forget that verbal grease accumulates about any word through meaning and connotation. My inverted commas are a plea to mind the semantics. But, of course, one must use such marks sparingly. L.S.

[2] Sir Arthur Eddington: *The Nature of the Physical World* (Cambridge University Press).

[3] Sir Arthur Eddington: *The Philosophy of Physical Science* (Macmillan).

"Wrapped about" and at times submerging the ineffaceable focus of awareness, our concentric chains of uncertainty screen a dual vortex.

Its one aspect is the ever boiling, ever changing and crumbling, ever dying and rising outer world – the *phenomenal* world, in the term of traditional philosophy. This world is a fata morgana spun of space, dimensions, directions, time.

The inner vortex is a calm "wave." Its nature we sense – even identify ourselves with, only at exceptional moments – as aimless, changeless, subliminal. Center of awareness or the *major-self*, this Ever Present is obscured by the flux and mirage of the phenomenal world. Yet it emerges in gleams of a peculiar vision.

We have evidence to unfold, and sense traces of the subliminal, even in the machinery of the material world: a ray of the *Real* in a mesh of happenings. With all its cracks and bubbles, the universe still is a breath of the *major self*.

We have reasons for a hope to arrive at the ultimate knowledge.

If, through the instrumentality of our eye and other tools of observation, we can go down to the innermost of our physical half, a minor shadow of the *All* ... if, through direct examination and a comparative analysis of otherwise observed psychic phenomena, we can go down to the rim of the second, the inner "shell" of our existence – the surfacial ego ... is it not possible, by a still subtler system of tests and probings, to obtain at least an inkling of the subliminal lining of our existence?

My task is to show that the instrumentality of music is spurred by that above mentioned peculiar vision; sparked in exceptional moments by the Major Self; and that the inner line of the musical order, that music's innate "design" reflects a subliminal gleam, that of the *Ever Present*.

The stream of things in itself we shall find to show a pattern hinting at the law of the subliminal – and one germane to the key order of music.

The universe of music presents then a life before and behind a unique "partition": one without dimensions and of unfathomable spirituality.

From "without", this partition – the immediate self – is stormed at and corroded by the torrent of the phenomenal world; from "within", it is stroked by the calm flow of the subliminal.

Two questions, if answered, allow us a forecast of, perhaps even a glance into the cosmic essence of music.

What does music take from the observable world? What, may we presume, music takes from the inexplainable, the subliminal?

The reply to the first query is: time, of course, to begin with; time, whose one facet is rhythm.

But the other basic elements of music's physical web are spun of the same thread as time itself. Tonal key and even harmonic chords are the same cross of co-existence and relatedness as is the time-moment – and indeed, the time dimension.

Thus, as an event of the phenomenal world, music hangs by the thread of the fourth dimension.

Being a ground mark of the empirical world, time is, by the same token, a source of our deception. A streak of awareness – or a "dimension" of awareness – time indeed is: not of the subliminal, the *Major Self*, but of the relative, bedazed, stumbling *minor I*, the immediate and personalized self. Time is the latter's way of stringing on, relating and binding into unity the wild dance of sense – data which make up the fata morgana of the phenomenal world.

This is why music, by its order a mirror of True Being, as I will try to show, a mirror that reveals the law and lining of the Only Existing, embraces also the deception of rhythm, physical flow, time. Uncertainties and blunders making up the "life" of the *micro-ego*, all have their traces in music.

What does music take from the subliminal world? In which way may music graze that world?

An answer to queries of even more limited aim is important. Why is a significant form the essence of art? Where and how does the role of art become a point in epistemology? How is a significant form related to the *major-self*? How does it differ from other gleams of self-transcending awareness and of creative seership of art?.

These queries will be answered, more or less. But one will have to bear in mind my point of distinction.

Epistemology is the study of knowledge in its rigid, outer state. Only metaphysical analysis carries the intimate, fluid, unfolding view of the very instrumentality of knowledge, of its rotating mirror.

The moment of closeness to the subliminal comes with the peculiar listening, with a release of energy at the high crest of intentness that marks the aesthetic catharsis. Our self is then freed from the "here-now"; is shorn of our basic deception and of the dream-blunders of the *minor* or limited *ego*.

In music this cosmic insight or intuition, eye of the subliminal vision, is awakened through the obsession of rhythm and form. I take "rhythm" not as mechanical time – derivative, but in a spiritual sense: that of order. Even when time-beat does transpire behind the flow of rhythm in such waves of the tonal universe, for example, as Schubert's song *Das Wandern* from the "*Schöne Müllerin*" cycle; or in the flow of Mozart's G-minor Symphony opening; or in Bethoven's finale theme of the violin concerto – they exhale the peculiar hypnotic power vested in their rhythm.

This is what loosens the seams of the phenomenal world and touches off the intuitive power of the *Self*.

Light is thus thrown on the metaphysical meaning of the *ostinato*.

Through the momentum of a persistent repeat, and for the duration of the aesthetic spark, our perceptual drive is dimmed and the dream-blunder of the outer world with it. This allows us to immerse our *I* into the air of the True World and sense its order.

The magic persistence of tone wave in Schubert's *Wandern*, or in the scherzo of Bethoven's *Seventh*, speaks of the cosmic pattern as eloquently as Heraclites' "*panta rei*" ("everything streams"). It starts the current of intuition listening to the whirr of the subliminal.

Where the marvel of music precisely dwells, is in its being the *only absolute art*. The very freedom from separate blocks of existence, trait of the empirical world, singles music out as a keen instruments for the sensing of the subliminal.

Of all the arts, music alone makes us aware of the mirage of individual separateness which marks us as units dwelling one "next" to another. Far more than any other art, music causes the *minor ego* (the empirical subject, in Kantian term) to fade at the showing of art's power and give way to the sense of emerging into the *All*.

This explains, among other things, the peculiar power of folksong. There is a high sense of tonal and spiritual latitude evoked by folksong; it comes from the first turn in the liberation from *principio individuationis*. This is also the source of a sharply felt significance and richness of personalized music conditioned by folk-chant.

In the delicate yet clearly present awareness of deception by the empirical world; in the vanishing of the personal and of the separate – a vaning effected by music; in the understanding of the ground error of the immediate *I*, arena of the phenomenal world; in the sensing of the timelessness and pointlessness of True Being – speak the glints of ontologocal awareness.

Persistence of repeat in music, its *ostinato* of rhythm and obsession of form, signify in some way the reappearing gleam of the aware *major self*.

Half-intuitive perceptions, they belong to music in exactly the same way as the corresponding discursive ideas belong to philosophy.

From now on – added to allusions to subliminal in music – my discussion will embrace these matters:

Evidence presented by cosmological abstractions and generalizations.

Evidence presented by the physical and the still deeper soil of the tonal universe.

Comparative view of the essence of logic and music; their place in the vision of the *Ultimate*.

Elliptic time, music and my trivium.

Ontological order of music and the meaning of *Return*.

II

Now, to those of our sense-data that shape into patterns of outer events where the subliminal seems to transpire: the glimmer we come to observe in music.

I wish to pass up obvious analogies: such, for example, as in Kepler's reforming of the old ptolemaic, that is, the geocentric view of celestial movements, and linking it to musical dogma of ancient Greece. Kepler saw in the pattern of his day's Renaissance polyphony an affinity with his own heliocentric principle of celestial mechanics fathered by the Copernican age.

Indeed, I shall also leave at rest the popular analogies in cycles of sun spots, growth of trees, economic disturbances, tempting as it is to compare them with the principle of return ruling the flow of music. In either realm I shall look for subtler facts, for elusive lights liable to provide us with a deeper test of any general view of the universe.

Our first sheaf of evidence is hidden in something entirely feasible or even demonstrable – through motion picture technique and electronics.

Time is reversible, and so is the curved clef ∾ of "cause and effect," a relationship which our mind has grown accustomed to take for granted.

But note this. The 'cause and effect" clef ∾ embodies – even visually, if we accept the symbol – co-presence of a kind *plus* functional relationship. This is what "cause and effect" really mean.

The famed Oxford scholar, Dr. E. A. Milne has stated the difference between the physiological or private time (that of the individual sensorium) and the nebular time, dynamic and speeding up, conditioned on the light from stars that may not even exist any longer: the ones that generated light hundreds of millions years ago and hundred of millions miles away. The rays from such stars, even when they had already exploded, may be reaching the terrestrial observer now.

But we also can imagine, even if we cannot figure out exactly, a nebular situation in which an observer on one planet sees events in time flowing in one direction, and another observer with a different mentality living elsewhere, following "at the same time" [1] the same line of events running in reverse. This amounts to the springheads of two furrows of time flowing into opposite "directions" that may co-exist and meet at some point of the curve of events. At any rate, the springhead of a line of time-events has no absolute "beginning" in cosmic time – precisely as a given point in a circle is never its "beginning."

In mathematics, one could suggest a counterpart to a curve of time-events by its nature – with no beginning. I am thinking of a concept used by the theory of polygenic (complex) functions: a circle in which one point is omitted.

In electronics one finds a spectacular reversal of cause and effect "explained" as an advanced potential. Electron B stirs even *before* the motion of electron A "causes" B to move.

But in the same way as the time-groove or the "cause and effect" clef ∾ can be reversed and "point of departure" shifted, the musical circle of theme and variations can be reversed and made to start its flow at any given point. I say "circle," because the form of theme and variations is but a foreshortened "theme-variations-theme." Here the return is one anticipated, not enacted.

Music knows examples of theme and variations reversing the usual order: *starting* with variations and coming to the theme, actually stated or anticipated.

It is the working of the same *Mind* or *Self* that shapes and adjusts the grand lines of cosmic data, exactly as an architect gathers mounds of building material at various points of his steel frame and girders. Rotating the curve-line of theme and variations in music and choosing its starting point at will, come from the same *vision* that fashions the various frames of space and "arrows" of time; that is, strings the sense-data into a coherent view of the physical world. It is the same mind of the cosmos indifferent to the direction of flow or time-order of events, ever ready to reverse them.

---

[1] Here my inverted commas are again a warning to mind both, the semantics and relativity.

Planck's quantum picture of Cosmos with its basic knot, constant *h* – one of the fundamental constants of the observable world, gives us still another glance at an order intimating an affinity between the subliminal and the dominion of music.

In the light of relativity squared with Planck's quanta in a later Einsteinean equation, energy (transformable from and into mass) is but an aspect of curvature in space-time, "frame" of our four-dimensional universe. Transformation of energy in itself, the limits of its reversibility given in the second law of thermodynamics ("the entropy of cosmos tends to the maximum") is another gleam of the subliminal order of Return. Entrophy is deadened energy lost in a machine turning heat into motion, as a part of energy becomes irreversible. The entropy of the physical universe "began" with the entire sum of power visibly transmutable, and must end with the entire sum of energy completely irreversible – even in an expanding universe; even in a seemingly limitless "explosion" of mass into energy.

This is the curve of a reversed return, akin to a specific toneform: the case of "variations *before* theme."

The fact that Cosmos is "indifferent" to direction either in space or in time; that a curve of happenings can start and effect a return at any point of any line of events, is a telling hint at the subliminal order and, as we shall soon see – also, at the rule of the musical flux.

The very cycle of universal constants, including Planck's *h* and the *cosmic lambda* (rate symbol of the expansion of the universe), has an inner and invariant numerical relation as the the cycle of conversion does. The latter reads: energy-matter-energy, a roundel in itself.

In the chain of uncertainties we belong to, its most cryptic facet is the unpredictable behavior of nature, the latter being but a web of observable relations and probabilities. We live "amidst" symbols of something not in itself real, that is to say, we live "surrounded" by squares of irreality (taking "square" as a mathematical term).

Are these odd leaps in the logic of nature so discouraging that even a clear minded and determined "critic of abstractions," such as as Ernest Nagel, poses the question:

"What *logical* compulsion is there that even if the world does possess a perfectly integrated logical structure, human thought should seek to encompass it?" [1]

The fact that the explosive, unforseen behavior of quanta of energy is tearing up even *statistical* law in the microphysical world – the law of probabilities – thus upsetting the pattern of events; and the fact that uniformity stops at the threshold of the infinitesimal, point to a mystery of the unseen. Do we not sense, perhaps, a trail of the *Major Mind*, of the indwelling "ultimate," privately, directly and "moodily" (which means, unexplainably) intruding in the machinery of the phenomenal world?

Do these fits of quanta explosions signify only mobile wrinkles on space-time? Are they not perhaps also glints of the *Infinite Mind* or *Major Self* which seems to manifest itself somehow in the general frame of even the phenomenal world, in its norms and patterns?

Discussing the impossibility of calculating either the past or the future velocity or energy, or path of particles – a situation already predicted before in the Heisenberg principle of uncertainty – Einstein showed the result of a simple experiment.

"The principles of quantum mechanics must involve an uncertainty in the description of past events which is analogous to the uncertainty in the prediction of future events." [2]

Uncertainty in the describing of past sense-data comes from limitation of knowledge derived from the measuring of momentum.

This startling fact, too, takes on the meaning which I have given above – and tentatively – to the fitfull action of quanta: still another, a very telling, feasibility – an "intrusion" of the subliminal into the ordered phenomenal world. The nature and logic and instrumentality of such intrusion *may* be found less cryptic than it seems, as we shall see in the closing section of this essay.

---

1 Ernest Nagel: *Sovereign Reason* (The Free Press). The word "*logical*" is italicised by me. L.S.

2 Albert Einstein: "Knowledge of Past and Future in Quantum Mechanics" (*Physical Review*).

Some stern "scientific philosophers," among them people who allow intuition to get quite deeply into the web of their science, will deem my *aperçu* naive if not worse: irresponsible. However, Dr. Percy Bridgeman of Harvard, famed physicist, is unbending in his insistence that the best in science comes from doing "one's damnedest with one's mind, no holds barred."

True, at one or another moment of its weaving, our higher mind is hampered by the basic, native and antropomorphic cast of our mentality, *plus* of course, all kind of semantic aberrations. It is revealing, however, that given such limited, clumsy mental apparatus, one with a wrongly set axis of observation and thought, we still manage to lift ourselves so high above our "terrestrial" mental level.

Well, not everything that *looks* anthropomorphic, *is* so. The suspected "intrusion" of the subliminal into relationships and "tangential" events of the outer experience with its "laws," and the entire film of uncertainty in which cosmos seems to be wrapped, gives credence to my above proposition, even if the terms in which the latter is couched, taste of the antropomorphic.

And, by the way, is not the thought of a firm, rounded unit of conciousness being but a fiction – as William James claimed – also leaving us with a stream of discontinuous, explosion-like *psychical* quanta? May not this fact be a symptom of the subliminal just as, in the final account, physical quanta may well be?

Here, however, emerge series pegging the run of civilization, to show that, in spite of William James' critique, one does perceive an ordered curve even in the world of moral and mental events.

Croce sketched the ethical pattern of Western civilization as a symphony on three motives: the creeds of liberty and intelligence *plus* the religion of Europe. [1] He did this not because of the contents of his Western ethics. Unknown to him, Western civilization's pattern was tracing a musical design. Croce's "ethical symphony," to him just a poetic *aperçu*, is in fact a modified ellipsis where the religious strain traces the circumference, with the three related motives centering themselves as the foci. In music this pattern is a rondo of the higher form.

---

[1] Benedetto Croce: *History of Europe in the XIXth Century.*

The cyclic contour covers still other facets of civilization such as economics, in themselves interrelated with ethics. [1]

The laws or curves of economics do not change; only the religion of economics, their driving wind, does. Before our own eyes the religion of competitive society is yielding to the creed of welfare state. Yet the perennial forces and trajectories of economics always return.

Coming back to ethical "symphonies" again – roundabout this time – we find ethics to be but points of aesthetic behaviour merged with and driving human dynamics. We come to understand the meaning of ethics as a code of social convenience and social aesthetics, as a receipt for co-habitation.

In this way the run of ethics and economics can be clearly traced as inclines in the surfacial curve of civilization.

Every pattern of humanity's basic thought, every vista in its concept of the world as a whole, is shaped in cyclic or elliptical curves: whether it is the ancient Greek logic of inclusion and subordination (one example of which is the classical syllogism with its visual groove in the Euler circles); or our time's organistic idea of things and events as rounded entities built of elements interacting internally. The last concept is, indeed, elliptic, with *n* foci allowed.

Yes, the same sparks of the intangible and of the unforseen hint at the subliminal *Ever Present* – in "spiritual" events, as they do in physical.

Coming back from and "*out*" into the macro-physical world, we perceive that its wave-picture and nuclei-of-energy picture, or the great law of thermodynamics [2], do not contradict the law of cycle.

And – in an astonishing bit of affinity – the shape of the atom itself is, according to all evidence, round or elliptic: shape of the

---

[1] My point apparently clashes – but really only on the surface – with a mind I so greatly revere: Dr. Morris Raphael Cohen. He censures Croce as well as Hegel, Schopenhauer, Bergson, William James and some other towering minds for their "monistic mania" in fixing "simple keys to philosophy of history." See *Morris R. Cohen's Philosophy of History* by Leonore Cohen Rosenfield (North Holland Publishing Co, Amsterdam).

[2] Cosmic entropy tends to the maximum.

atom, note it well, that of an infinitesimal pattern of a solar and cosmic order.

In this fitful, moody, change-hungry tangle of events of the physical world, only the "return" is sure.

But "return," the visible face of immanence, indwelling, ever-presence, is the law of music.

III

Let us, however, go down to the physical under-frame of music itself and then, look above – to its semantics. Let us see what can be read there that is beyond the tangible and the observable.

To begin with, the acoustical substratum of music lends it a kind of immanence and makes music a correlative of the universe in the same sense as mathematical norms are. In music, too, there is an eery residue of fictional reality or feasibility somehow related to the *norms of being*: the same as in purely mathematical symbols and allusions – infinity, constants, curved dimensions, imaginary numbers.

While being mere symbols, they surely are tangents of the universe since, in dealing with the "outer" world, science employs them forcefully and successfully.

A view of the acoustical sub-frame of music allows us to understand the nature and interrelation of music's basic elements: tone and rhythm and, as result – melody. The web of harmony, too, is lighted up in this way.

But even resetting music merely as a mechanism correlated with the physical world, one comes upon affinities with a breath of the "beyond" in them.

To start with, here is the kinship of keys or tonalities, each of them a minor world of related harmonies, and all of them – an order of coherence. We behold their stance of emergence: a circle within a circle. The inner ring of derivative natural tones with frequencies 2, 3, 4, ... times that of the fundamental sound, the vibration of which touches off the harmonics (third, fifth, and so on), is included in a wider circle of keys emerging in the order of tonal cousinship; growing more distant and linked to nascent overtones as tonics. The difference of "*temperament*" forces upon us a variety of such spirals of key-kinship.

Out of theoreticians' attempt to replace our tempered scale with a fraction-tone gamut of nineteen steps $(5 \times 3 + 2 \times 2)$ or even of thirty one steps $(5 \times 5 + 2 \times 3)$ – those comprising the natural sevenths –, a noted Danish scholar Thorwald Kornerup has unfolded the idea of acoustical relativity.[1] He even offers a constant for the transposing of adjustments within the tempered scale. This view seems to be a light in the same direction as my "curvature" in music's continuum of which I am about to speak.

The second aspect of tonal relationship looms in the gravitational energy released by the harmonic sequence known as modulation.

A unit of gravitational tonal force, the cadence, the essence of which is the triton (fourth + major seventh) with its resolution, bears a telling similarity to a nucleus of physical energy and its field. The triton and its resolution mark a definite position in the tonal universe even as gravitational and electro-magnetic tension in the physical world signify a change in the curvature of the space-time continuum. The triton f–b indicates a field of tension within C-major, while the inverted e♯–b means gravitation within F♯ major. By the same token, a circling increase of tension marks the swift narrowing of gap between keys of distant kinship through sudden modulation. [2] It is as if a gravitational force is released suddenly – to press even distantly related keys into affinity.

We could think of the following equations or – to evade dispute – analogies: key with its kernel, the triton, as corresponding to a frame of space; and modulation having the same meaning as a change of curvature of space. In Einstein's cosmology the latter is given as the meaning of gravitation.

---

[1] Thorwald Kornerup: *Das Goldene Tonsystem als Fundament der theoretischen Akustik* (Copenhagen, 1935).

[2] In the bushels of scholastic dissertations full of preciosity and muddle by that odd Viennese worthy, Heinrich Schenker, we find one of his rays of sagacity: the analysis of tonality. But fumbling with that, he still does not speak clearly of the functional aspect of harmony in relation to the cadence, and of tonality – in relation to musical form.

Extending my analogy, one could speak of polytonality, for instance, as the musical counterpart of the curved four dimensional space-time continuum and a function of four co-ordinates with the fourth multiplied by $\sqrt{-1}$ ct.[1]

I have shown earlier that gravitation is the essence of power released by a change of tonal key. Of course, it is gravitation of a spiritual, I am eager to say, subliminal nature.

Tonal balance plus modulation more than stress the power of form in music. They are the very lifeblood of form; play exactly the same role as the curvature of space, in the tracing of the silhouette of the *Whole*. It points to the way in which the unfolding Mind weans the physical universe.

Still more dynamic is the third aspect of tonal relatedness, the unfolding harmony and tonality.

The noted music scholar, a mind of high acumen controlling vast learning, Dr. Joseph Yasser, has given us a remarkable view of the ever-going rebirth of harmony; and he revealed the logic of this unfolding. For his doctrine Yasser chose the best possible testing ground, an ancient tonal civilization with roots still green, the Chinese. He compared the latter analytically with the oldest tonal cultures of early Europe so the entire spring of the world's tonal life is embraced.

Persuasively and forcefully Yasser presents the worldstream of harmony as being continually enriched, and also "pulverized," by the addition of auxiliary scale steps, fractions of tone, etc., with auxiliary steps growing into organic parts of the scale. Fusing the historic musical vista with Hegelian dialectics not always visible, Yasser shows how the main concept of harmony ingrown in man's tonal mind had changed its basis from the pentatonic scale (in Yasser's terminology, the infradiatonic scale) of the early Greek, Chinese, Hebrew, Celtic culture; how by incorporating the auxiliary steps the tonal world pattern had developed into the synthesis of the diatonic; how by a further antithetic process more auxiliary semi-tones garnered from the circle of natural harmonics, were first opposed to, then incorporated into the scale as its full-fledged steps – to form the atonal

---

1 The four-dimentional continuum. See Albert Einstein: *Theory of Relativity*.

scale, with the synthesis of supra-atonality and still higher and just as natural complications to come. [1]

In this logic of unfolding harmony one finds what one has *a priori* looked for. First, a universe of tonal confines and cosmic "border-lines" implied in the finite physical world of our day's cosmology. One perceives something else, too: a glance into the subliminal, as the squaring of all evidence will show. Yasser's – and the tonal universum's – evolving harmony can be compressed into an image of symbolic logic: a variation of Euler circles' with new contents.

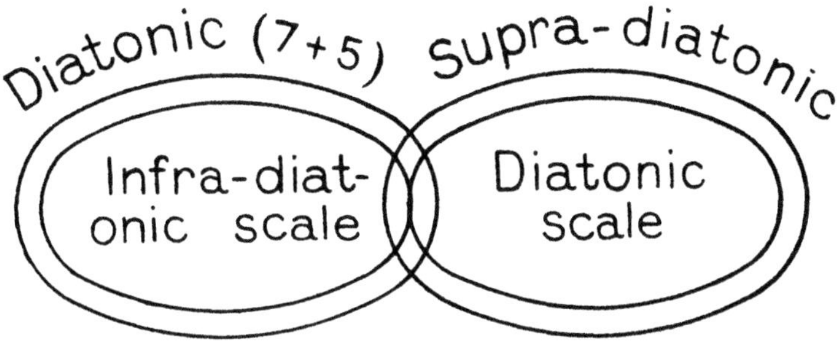

– then more dual circles: the atonal scale enclaving the supra-diatonic; the supra-atonal with the atonal – *n* overlapping links of them, and so on.

Dissecting the physical body of music allows us to observe the trail left by a musical form in action.

Another noted scholar of our day has given us a lucid summary of such an observation.

"Before a symphony can be played by an orchestra, there must be collaboration of many parties – a composer, the makers and the players of many instruments and the conductor of the orchestra. All are, or have been, at work to produce – just a curve ... All the art, all the mannerisms, all the successes and failures of these many workers are embodied in one single curve.

---

[1] Joseph Yasser: *A Theory of Evolving Tonality* (Chapter VIII), American Library of Musicology, New York, 1932.

When a gramophone record is made of the performance of the symphony, this curve is preserved in a tangible form." [1]

This means that the visible, the phenomenal existence in a stretch of music reduces itself to a curve which is an aspect of cycle, or rather, an anticipated cycle. For while curve is a direction, cycle is an order.

Of course I am speaking here of the philosophical meaning of the curve and cycle, not of mathematical terms.

Set this thought against the fact that the curves of frequencies left by the playing of a great pianist or a mediocrity pounding on the piano, or even a weight dropped on the keys, is the same.[2] Also note: what distinguishes music from noise, is the periodicity of the sound curve marking a musical tone.

Together, all above facts tell us this. The differential of physics of music and aesthetics of music can not be registered by a physical instrument and measured by an observer. Only a concept related to the spiritual and the subliminal can explain it. Secondly, what distinguishes music from noise, is periodicity or "return," i.e., the sheen of duration, of everpresence.

Music is alive not "because it moves and acts like living matter." Art's rationality is of another order.

The opinion of outstanding philosophers, composers and music scholars deems music an emotional release; both, a form of and an outlet for self-expression. But this view still does not cover the rationality of music.

In her fine book giving much thought to the semantics of music, Dr. Suzanne Langner states justly that "sheer self-expression requires no artistic form ... the laws of emotional catharsis are natural laws, not artistic ... if music has any significance, it is semantic, not symptomatic." [3]

Besides, the so-called emotional effect, so often mistaken for the emotional catharsis or for the "contents" of music, is but an after-effect, or a mere reaction to rhythm and sound. Our music, more particularly – its formal and polyphonic weaving, is nothing emotionally in comparison with the wild Indian outcry, the

---

[1] Sir James Jeans: *Science and Music* (Macmillan).
[2] Jeans, *ibidem.*
[3] Suzanne K. Langner: *Philosophy in a New Key* (Harvard University Press).

pitch and ecstasy of a Galician hasidic dance tune (that of Jewish pietists), or the frenetic skirl of a Highland fling.

As for moral symbols touching off musical concepts or emotions, they are but *one* case of ignition; not more important than any other spark igniting an art-form. Beethoven's *Ninth Symphony* is a telling example of a not perfect, yet significant form fired by a moral spark. But the rationality of Beethoven's *Ninth* is not covered by the weight of its emotional or moral catharsis.

Neither is it covered by a synthesis of physical and physiological patterns with which a new music "science" tries to compose significant musical forms at will.

A picaresque figure in music, neither quite a creator, nor quite a scholar, had spent a prodigious amount of work and of spectacular, if haphazard, documentation-to prove that musical science *alone* can create any form of music.

"If art implies selectivity, skill and organization, ascertainable principles must underly it. Once such principles are discovered and formulated, works of art may be produced by scientific synthesis." [1]

This is a clear statement, plausible on the face of it. It hides, however, fatal blemishes. First of all, "selectivity, skill and organization" do not cover all that transpires in a significant form. Secondly, the above dimensions of art are not absolute. By the nature of things, "selectivity, skill and organization" are related to, and can be applied to only special generative ideas, to motives peculiar to concrete craft.

Art's logic of the "whole," its rationality strongly differs from that of discursive reasoning and measuring which is the heart of "scientific synthesis." The differential of physics of music and aesthetics of music can not be made good in the workshop, by calculation, measuring or shifting of physical patterns.

[1] Joseph Schillinger: *The Mathematical Basis of the Arts* (Philosophical Library, New York), a curious work by a man whose brain was strong, philosophical training nil, scientific-negligible; whose fine gift of correlation tried at times to master a wild ocean of things hopelessly unrelated.

IV With these facts of the observable world in mind – physical, moral, semantic – we can at last glance into the inner fabric of music. Can find looming, there too, the silhouette of the *Beyond*, even if we are not granted a precise defining of what music may draw from the subliminal world.

If rhythm, an underpinning of the phenomenal world and, by the same token, of concrete, physical music, can yet, as I have shown, touch off a spark in the subliminal, the other aspect of aesthetic obsession, musical form, can be shown to be the very ray of True Existence, through music's earmarks of pointlessness, duration, universality.

There is a structural kinship – one stronger even than that of rhythm, a kinship which music draws from the "beyond." The obsessive power of rhythm splitting our self and releasing its hidden ray of subliminal vision, can be observed in a still more pointed way in musical form. We come to understand music's inner necessity for the "return."

Music's matrix, or its equation reads: A + B + A = Ring = Cycle = Return.

Either in a minor tonal structure (motive A *plus* motive B, or transition, *plus* motive A); or in a major frame such as the sonata (A, exposition of themes; *plus* B, development; *plus* A, restatement of themes) – the cycle looms in a direct or in a veiled outline.

Here is the law ruling the physical wave of music; the order of music's phenomenal tides storming the "partition" of our immediate ego.

In this music is seen and is acting exactly as the general stream of physical events does: as a ring – *annus* or *annulus*. The substantive musical idea or motive has the urge to return as much as the seasons of the year (*annus*), as the tides, the thirst for mating in living things, the peak of harvests.

Taken in a spiritual sense, the principle of Return inherent in pointlessness and duration or ever-presence, is also the order of the True Inner World "behind" the immediate ego. In our divine moments we sense it.

It is revealing to find that even mutations in the cyclic pattern of music correspond to changes and extensions in the circling lines of the physical world.

Let us note this amazing grouping of facts.

The principle of Cycle transpires, among others, in the lower rondo forms (first and second) or in the sonata form of music – just as it is present in the orbits of satellites, the moons.

The modified circling form, one with two foci – the ellipsis – corresponds to rondo of the third form in music: A + B + A + C + A. Here two new motives are interpolated in the flow of the recurrent first motive A. In the physical form of ellipsis the circumference corresponds to the flow and return of the main motive A; and the two foci – to the new motives B and C that underline the constant A. But observe how precisely the orbit of planets follow this modified pattern of musical thought. All at once we sense it strongly: the two orders *must* flow from the same ultimate spring. The principle of *annus* we are bound to interpret as a gleam of the Major Self, the *One* and *All*. Thus we find again that the basic order of music suggests the subliminal facet of Being; mirrors the interplay of the major, All-Aware Self with the worlds, "objective" and foreseeable.

So native is the sense of ultimate oneness to music that a grafting of musical pattern on other art-forms heightens the feeling of universality and immanence. At times, such a grafting gives art-forms other than musical a certain cosmic width.

In rare cases only is a poetic order fashioned – subconciously – after musical form. Such, for example, is Pushkin's *Georgian Song* where the return of the first image gives peculiar point to the personal emotion: adds to the tension of the empirical ego a breath of the subliminal.

As in Pushkin's case, a vision of genius made Stendhal think of a cyclic play: one to take place on a circular moving stage and by rotation show in precise repetition – a man's thought, feelings and deeds of one day's life. The wheeling stage meant that Stendhal, perhaps only instinctively, responded to our ever-hidden urge of return, to the musical urge of the rondo.

An ancient book of Chinese temples, that wondrous, five thousand year old Yih-King, is built as a rondo of the higher form. Its basic part is enclosed in a variation of the circle: a hexagram connoting the web of life with the inner great circle meaning the universal.

An outstanding American musician, Dr. Mark Brunswick, toiled nobly to prove that the sonata form is "evolved out of the exclusive musical dimension of tonality and has no sense or even existence apart from the art of music.' [1] While the first half of this statement can hardly be challenged, one is not sure of the second.

Tonality and its kernel, the cadence, are indeed sourced from the cosmic order of life – the Cycle.[2] And in turn, cadence and tonality are the driving force of the sonata form.

But coming to still another unusual literary pattern, the "overture" in *Swann's Way* of Marcel Proust, we find it to be a variation of ellipsis in its outline, and a rondo of the higher form in its structure, filling the formula A + B + A + C ... + A. The basic and recurring motive A is the dream-image of "*mamma.*" The others are: Combrai, country place where the narrator visited his aristocratic friends; Swann himself, etc. In its outline, the "overture" is thus elliptic: the recurrent image of "mamma" propelled by a circling flux of semidreams, is the circumference; and Combrai, Swann, etc. – the foci.

This form comes very close to the sonata pattern: development in lieu of, or together with – secondary motives, does have its counterpart in this literary pattern.

Thus evanescent mutations in our psyche half-asleep flow to firm cadences. This flow hides a taut order revealed in art. The world of mental twilight and dream conforms to the cyclic – and cosmic – pattern of music.

Studying the above pages of Pushkin and Proust still closer, one is forced, however, to grant little weight to a *surfacial* sameness of material or form. Even with the tonal sisters – poetry and music – their similarity can be deceptive. For in either, poetry or music, the real material is not sound (this is only

---

[1] Mark Brunswick: "Tonality and Perspective" (*Musical Quarterly*, October 1949).

[2] Of course, I use again "cycle" not as a strictly mathematical term but as a philosophical concept.

medium of transmission), but implications of the sonorous flow: images and symbols intimated by the phenomenal world in the first; and patterns abstracted from the stream of sonorities in a *wholly different* way and shaped into forms that properly belong to music alone, in the second.

At any rate, the casual kinship of form in even outwardly related arts, that of the word and that of the sound, has a deeper spring than physical material or the blared notion of symmetry. We shall see later where the glints of affinity come from.

But let us return to Stendhal for a moment. We perceive that his play-photo of a man's emotional life and action of a day's duration thrown on a wheeling stage, could be either a circle or an ellipsis depending on a number of basic motives around which the action centers. In its musical meaning the shape would be correspondingly a rondo of the lower or higher form.

I feel, I must quote from yet another scroll of creation to sustain my point: the form of my own *Third Symphony* ("Symphony of the Seas").[1] This form shows a structural tonal pattern visibly reminding Stendhal's scheme. At a closer glance, the form of my symphony is identical with the frame of Proust's "overture" in *Swann's Way.* [2]

In that *Third Symphony* of mine, the form seems unusual but only at the first glance. The two sections, sonata and intermezzo – finale, are linked in an elliptic outline. The movement of the main theme fully returning at the end of the second section, creates the circumference of an ellipsoid, and subsidiary motives – the foci.

The symphony's physical contour corresponds to such higher

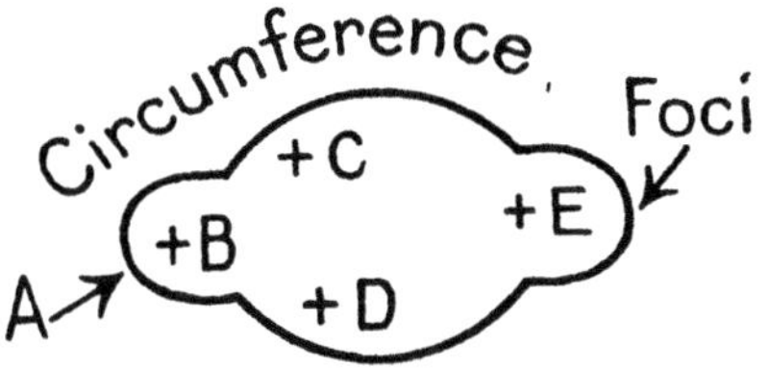

[1] Published by the Universal Edition in Vienna; European première by the Colonne Orchestra in Paris in 1925; American première by the New York Symphony in 1927.

[2] At the time of composing my *Third Symphony* I had not heard of even the name of Marcel Proust. L.S.

roundel trims as the third and still more intricate ones, with three or more subsidiary motives: meaning a modified elliptic silhouette.

Though not programmatic, the spirit of this symphony is revealed dually: through its tonal substance and also, in my own poetic motto of the piece:

I am a vibrant atoll,
In sea noises I am bathing,
I live on the greedy cries,
Rush after the voices.
Far away
Glitters and murmurs great Atma;
Enchanting the murmur,
Enticing the tale of seas. [1]

For a musical piece driven by the sense of communion with Cosmos, why have I chosen this very form? And why this theme or pattern bound to carry in its potential the form of the whole, as themes do?

I had taken this form as any composer does, without being concious of it *as an individual*: because of a cosmic emotion and the prompting of the Major Self, source of creative and of subliminal vision. The Infinite I, the *Universal Self* – is what speaks in music so forcefully!

Of the unusual variants of the cyclic form in music that find their counterpart in poetry, one is Vincent d'Indy's *Istar* for orchestra. It keeps the music unified and definite through a line of thematic mutations, not by a bold stating of the main theme. If we consider the last showing of motive only as the *final variation* of theme, not yet the theme itself, we may insist that the main theme is never heard. At any rate, such a musical form, even if only suggested here and not yet materialized, is logical and should have all the binding potency.

As it is, *Istar* is a cycle where the tension of return is effected by hints at the basic thought wrapped in a mantle of variations. The form of *Istar* is thus suggestive of the "contours" of the subliminal world as sensed by the Major Self: a spiritual pattern transpiring in variant silhouettes of the empirical universe.

[1] Quoted from Lillian Saminsky: *Poems and Adaptations*, published by the Oxonian Press in Oxford, England in 1931; second, memorial edition in New York, 1947.

Again, my thesis about the native order of music as the mirror of Being seems sustained by a salient art form.

In his book *The Poetry of Edwin Arlington Robinson* the American writer Lloyd Morris shows that Robinson's poem *The Whip* repeats the form of *Istar*, the main theme never appearing save in variations.

> The doubt you fought so long,
> The cynic net you cast,
> The tyranny, the wrong,
> The ruin, they are past;
> And here you are at last,
> Your blood no longer vexed,
> The coffin has you fast,
> The clod will have you next.[1]

But there is still another, a most unusual example, now taken from plastic art, that presents the ideal outline of tonal ellipsis.

Alfeo Faggi, an American of old Florentine lineage and one of the greatest sculptors of our time, has created for the chapel of Chicago University bronze doors whose style is our day's metamorphosis of the great Ghiberti manner. I mean, of course the Baptistery opposite the cathedral of Santa Maria del Fiore, or the Duomo, in Florence.

Formally, Faggi's *Saint Francis* is nothing less than a strictly musical mould known as the rondo of the third form, mathematically expressed as $A + B + A + C + A$, where A is the thrice quoted main subject. It is curious to find that within his form Faggi uses the dialectic development of his material, a typical proceeding in musical thought. His subordinate motives struggle with the main subject through sheer dynamic variance. A telling play of difference of potentials as classical mechanics have it!

The main theme A is the image of Saint Francis, a cadence of loving humility and contemplation. One sees it represented in the first panel (Saint Francis in meditation); also in the third (the same state of spirit but in the milieu of wilderness); and in the fifth and sixth forming a continuity: Saint Francis listening to the birds and praying on the mountain of La Verna.

---

1 "The Whip," opening stanza and first variation on an invisible theme. Edwin Arlington Robinson: *Collected Poems* (Macmillan).

Our *B* and *C* of the rondo form, the second and fourth panels (young Francesco Bernardone on horseback starting on his path of humility and haranguing the heavy-minded, resisting peasants) are transitional elements of accentuated movement which counterbalance the peace and transparence of the main subject.

The five monks over the tomb of Saint Frances form the coda of this extroardinary rondo. Its crispness, grace, telling silences –are a marvel of our day's art.

One could run through a wild variety of facts of art in which the universal principle of *annulus* shines through a pattern of mobile symmetry.

Here, to begin with, is the Greek axiom for dramatic unity reframed with amusing plainness by Tchekhov when he said: "If, in the first act of your drama, a rifle hangs on the wall, it must shoot in the last."

There is the *gnomon*, the forgotten ratio of dynamic symmetry, known to ancient Greece and artists of the Renaissance as "the golden section"; used by Albrecht Duerer and da Vinci: the *gnomon* which apportionned and rotated its sections, angles, designs even as the sonata form does.

One thinks at once of dynamic inversions of a musical form: making its pattern run backward; turning it upside down; shifting the cells and motives within the formal frame; exhausting all the variants of a mobile symmetry.

But ... is it not revealing to come across an Einsteinian *aperçu* that allows us again to marvel at the cosmic pattern of return lurking in unpredictable corners? In his *Ether and the Theory of Relativity* [1] Einstein discards a theoretical construction *because of its assymetry*. It is incompatible, he says, with the conformity of physical data which the theory tried to unite.

I have just said that in art design the *gnomon* rules the

[1] Einstein's re-edited address at the University of Leyden.

rotation of circumferential patterns – circular, elliptic, rectangular, octogonal – and of their foci, even as a sonata does.

This datum hides an infinitely deeper meaning than that of a point in the explaining of a distinctly musical form.

Watching the spread of the *gnomon* domain, one discovers a mathematical ratio in the nature of a cosmic constant to be the key to a strictly musical pattern.

Research of the last decades has shown this.

Formal patterns seen in things of nature – in stellar galaxies and orbits of some planets, shells, leaves, discs of plants, human bodies – or in works of great painters and builders (in the ancient basilicas of Southern Italy, for instance, or even in the pyramid of Cheops) – are found to be ruled by the golden section. And this principle is defined mathematically as a case of the Fibonacci series: $\psi^1 + \psi^{n+1} = \psi^{n+2}$. The golden section rule's geometrical counterpart is the logarythmic spiral: an equiangular, flat curve rolled on itself *ad infinitum*, grimly following the increase given in the Fibonacci progression.

First studied by Descartes and named by Jacques Bernouilli of the famed family of mathematicians, curves of this kind were also singled out by Albrecht Dürer who called them *muschellini*, shell-lines. They affected Dürer's delicate curves. The great painter sensed the order of life and growth exactly as the Ionic architect did when he fashionned the leaf and curve – design on the apex of his exquisite column.

But here was a fact new to mathematicians and philosophers. The rule of the golden section and of equiangular design can be traced in tonal forms, too.

More recently, analysis of the role of the golden section in poetry and music may have been ignited by a forgotten observation of a century ago exhumed by music scholars of our time. In many cantos of his *Divine Comedy* Dante used – unwittingly, no doubt – the "divine proportion." [1] The *gnomon* was thus called by artists of the Early Renaissance when they were figuring out their design of human bodies.

Recent discoveries are, however, much more striking and abundant.

The structure of some of the ancient folk chants as well as of

[1] A. Zeising: *Neue Lehre von den Proportionen* (1854).

many creations of great poets in later days, Schiller's *Goblet*, for example, are following the golden section ratio. And so do the forms of Bach's *Chromatic Fantasy and Fugue* and of Leporello's *Notte e giorno faticar* in Mozart's *Don Juan*.[1]

Still more significant, a very recent study of the sonata form has revealed that its basis is "the golden mean form" – in classical works such as Mozart's sonata movements, and pointedly – in Beethoven's *allegro* of the C-Minor quartet, opus 18; that this pattern shows "fairly true through Western music" up to our own day, in works of Debussy, Schöenberg and Bartok.[2]

All of which is, by the way, quite a case against Dr. Brunswick's insisting that "the sonata form has no sense or even existence apart from the art of music," a statement about which I have cast my doubts earlier, but for other reasons: analytical as well as philosophical.

And now, we behold something of far wider import. The *gnomon* or golden section rule reveals to us – *and thus supports my thesis* – that the curves of nature showing in the physical aspect of art tend to only approximate the Cycle, never to close it.

The life curves tend to approach the "circumference," are perhaps *symptoms* of the subliminal: those of pointlessness and immanence. They do not match completely, in nature or in the physical medium of art, the *anticipated* "profile" of the subliminal.

Of this digression one finds telling traces in music. It is enlightening, for instance, to follow a dispute of two towering musicians of our time.

Arnold Schöenberg pointed out that in the tabulation of basic tones and their overtones "the six initial partials repeated at three different levels from the fundamental tones *c*, *f* and *g*, produce the acoustically natural basis of the diatonic scale, whereas an addition of the remaining seven partials to the same three-fold series produces the acoustically natural basis of the chromatic scale." [3]

---

[1] E. K. Rosenov: *The Law of the Golden Section in Poetry and Music* (Work of the State Institute of Musical Science, Moscow 1925).

[2] J. H. Douglas Webster: "Golden Mean Form in Music" (*Music and Letters*, July 1950).

[3] Joseph Yasser: "A Letter from Schönberg" (*Journal of American Musicology*). Dr. Yasser quotes Schönberg's ideas first unfolded in an address at the Berlin Academy of Arts decades ago.

Let us look at the similar process of tonal unfolding in ancient times, at the pentatonic scale common to peoples who had no visible historical bonds, who were extremely remote from each other – geographically, historically and culturally: the ancient Chinese, Jews, Greeks, Scots. The fact of a closeness of their melodic taste and tonal perception can point to only one thing: common roots of mental and melodic ancestry.

But following the same route of thinking, the deviation of pitch increasing in ascending overtones, which, as Dr. Yasser shows, viciates Schöenberg's table, can be attributed to those expected differentials of tonal perception or even of sheer biological hearing: to individual deviation which nature grants even its closest related creatures.

The cycle of overtones catching up with the repeated initial partials and closing their circuits – now and then, yet not fully, owing to acoustical deviations – present the same evidence as Dürer's *muschellini*. Nature only intimates the "return"; does not fully effect it.

Coming back to the life curves one must affirm still stronger, that while in their physical aspect they come close to circumferential, circular or quasi-circular lines (elliptical, rectangular, octagonal) – as they do in equiangular spirals of the shell forms or in the sonata pattern – they never *fully* mantle themselves in a subliminal "silhouette."

Like symbols clasping waves of the objective world in a formula, concrete art forms only hint at the subliminal, give us only an inkling of its anticipated essence: never fully close the chasm between the phenomenal and the Beyond.

Logarythmic spiral vis-a-vis sonata tightens the round of evidence coming from the concrete universe and its relation to art symbols. This ensemble grant us, at least, a premonition of the true order of Being.

V The law of musical form rises in springs much deeper than the aesthetic medium of the arts.

Faggi's bronze doors at Chicago University present a lesson of wide import.

Here a form strictly musical suddenly appears in all its technical armor from the depth of creative perception and in an art that is – in pith and order – so alien to music: in plastic art. A fact of prime significance!

The rule of return and of oneness in aesthetic perception speaks in many more veiled phenomena than we suspect. In the end, the unity of the arts asserts itself, but only in its remote aspect.

Viewed horizontally and statically, the unfolded cycle seems to speak in terms of architectural symmetry. Yet it shows its highest tension as a strictly musical order.

How do we, however, reconcile these findings with the fact that the material core of music, of sound itself, is but an uncertainty, or worse – a deception?

How can we even suppose that the visible pattern of tonal thought – ring or cycle – is in some way related to the essence of the subliminal world; that squared with the *immediately given* in our aware self, this observable silhouette looms like a gleam of Absolute Being?

"Ring" or "return" is indeed *our* notion: in its subjective aspect – a raw symbol summing up the experience of *our* senses. Yet in its ideal aspect, Ring, Cycle, Return, mean something prior to symbol, something pointing to a more lasting spring.

An affinity between the key form of the tonal universe and a key concept of mind persisting through the ages invite firm conclusions.

How significant it is to see one of the basic religious symbols of humanity – and one of the oldest – the *mandala* ("circle" in Sanscrit) interrelated with the key form of music, and also, with symbols of the outer world processes in which one senses a breath of the Beyond!

Variants of *mandala*, again a rondo pattern dominating the collective and submerged mind of humanity since the dawn of time, are the wheel, the cross, the flower, even the medieval image of the dragon swallowing his tail in a circle.

Taking the first affinity of the two worlds that occurs to me, I think of one of the marks of the physical universe, the entropy, as embodying the cycle in reverse. Of this I spoke earlier. If presented as a version of *mandala*, the entropy of the universe could be symbolized as a dragon breathing fire, losing heat and swallowing his fast cooling tail in a circle. In music it would mean a rondo with subsidiary motives passed first, and the main theme stated and reaffirmed later.

The *Real World*, the nature of which seems to be mirrored or symbolized in music's *mandala* – in its tonal cycle, appears then as a triune knot of pointlessness, duration, return: all of these – facets of the Eternal Round of Awareness, or the unobservable Spring of All Appearances.

This *Real World* is the invisible vortex "storming" the partition of the Major Self, everlastingly merging with the Major Self, the true and only Ever Present.

In its dynamic and emotive aspect music is an act of symbolic will, just as words are deeds of symbolic action.

Music, however, *only seems* to be a wave of dynamic power. It really is only a mirror, only an awareness of the "vortex" and of its "beyond."

Plato stated the core of this idea in words as eloquent as they were prophetic:

"Earthly music is only a shadow of true music."

Music's instrumentality is the eye that sees the Whole; sees the *constant of the True Megacosm.* Music is the vehicle for the "image" of the True, the nature or "dimensions" of which we do not know: subliminal dimensions.

Thus the essence of the musical universe reveals a peculiar facet of Being, where the change in variables (twists of the tonal mind, such as modulations) shows relativity, while the "pattern" is anchored to a subliminal "structural" order.

Something, someone, a continuant or spiritual constant of the *True World*, the latter's higher order, is playing with the outer universe, that is, with its own mirage. It is as if the Aware Self were holding a wry mirror to its own "face" and juggling false reflections.

Driven by the power of life and the flux of becoming, this mode of intuition in the Aware Self now plays with, then effaces the outer world; is now affected and deceived by it; then freed from it. In action and through a peculiar "hypnotic" effect – the latter in itself beyond our perceptual awareness – music straightens out the deluded self and makes it divine the One.

In the dual aspect of musical form, its physical foreground is under the rule of tonal ring; its aesthetical half bows to what we sense and try to describe as the cyclic spirit of the subliminal. Thus transpire the concentric chains of uncertainly in which we live. Transpire first and dissolve themselves in the Major Self, in the end.

My summary of the evidence, derived from the assaying of the objective world's fluxes and of the symbols correlating and systematizing our sense-data, has already shown this.

The outline, relationship of things and the threads of sequence in the physical universe are not mere phantasmagoria or random dreams of the Aware Self. In some way unknown to us – though sensed through intuition, in metaphysics, art, religion – even the shadowy outline of the observable world is a symptom of the subliminal.

Somehow the mould and lay of the observable world reflect the "silhouette" of the Only Existent: loom as its tangents. Even the seeming casualness of the phenomenal world, its very uncertainties hide a contour or a "method."

It could not be otherwise. For the phenomenal world is *functionally*, even if not transcendentally, an aspect of Being.

This in why, in spite of the moody behavior of the quanta; in spite of schools holding relativity to be not an ideal tool to deal with an expanding universe and trying to supplant – or perhaps, only supplement – the relativity principle with continuous creation;[1] in spite of all this, there is law and harmony in the physical world. The cycle of its constants fundamentally related[2], is a telling show of our point. In the outer world its

[1] Fred Hoyle run over by Bergson (in *The Nature of the Universe*).

[2] The speed of light; the mass of the proton; the mass of the electron; the unit of electrical charge; Planck's constant (h); the gravitational constant; the cosmic constant (*lambda*).

harmony in itself intimates the existence of the *Beyond* with its own order.

Thus, if the inner pattern of the world's main physical process is a cycle ...

If the inner pattern of revealing acts of art is a cycle ...

If the inner pattern of humanity's history is a cycle ...

If the inner patterns of moral history and civilization are but variations of cycle and return ...

If the inner pattern of life of the Self is a cycle ... – does not this compel us to think of the cycle as the indwelling, we presume even, the subliminal, order of Being?

Fact being "something we find, and fiction – something we create," the Cycle remains clearly the mode we see everywhere: in essences concrete and in media intangible. In a superior and most potent meaning, Cycle is then a fact.

. . . .

All that has been stated above is not mere discourse drawn from a new metaphysical grammar, applied – in a light-minded analysis – to the "oozy and transitory stuff of concrete evidence."

It is true that a new metaphysical grammar armed with mathematical logic and derived from an utter refining, or may I call it "skeletonizing" – of perceptual evidence, does press the latter into stiff moulds of reasoning. But when we confront this evidence with logic seen in another light – as a closely clipped, unified view of mental confines grazing cosmic probabilities – the whole picture changes.

To begin with, a condensing of concrete evidence into symbol and language is not even indispensable to knowledge. Language is only a codification of symbols, not an instinct, as animal psychology shows. Blind or deaf-born or apes develope judgement without forming a language of their own.

On the other hand, our abstract thought is capable of deliberately passing perceptual evidence, of acting in spite of it. At the highest point of such awareness, mind visions an entire universum of feasibilities. On another level, our thought straightens out, codifies, even shapes perceptual evidence in accordance with an anticipated principle, as masters of mathematical cosmology do: Einstein, Planck, Eddington, Milne.

But indeed, logic is not a code for corsetted reality. It is an infinitely wider reality.

In this light, logic of the *highest* level is revealingly akin to music: to the latter's foresight or forecast; to its detaching itself from the physical; to its *kind* of dematerialization; to its kind of marshalling implications and potentials.

For in their deeper meaning both logic and music are a vision of spiritual contingencies.

When the ancient Greek mind, designing its *quadrivium* of sciences decreed for a thinker, placed music next to mathematics,[1] it was prompted by a deeper thought than the Greeks were aware of. But we see, there is still another sister: logic.

Kinship of the three lies in the essential instrumentality common to members of this, *my own trivium.* The instrumentality native to logic, mathematics and music dwells in their vision of the immense film of the feasible.

An imaginary number, $\sqrt{-1}$, for example, is not a sense datum, nor trait of an object, but a tangent or mode of a feasible variation in an existing thing: a twist of relational logic. But note that the same order of existence is that of music. The latter's inner being is not a sense datum; its inner pattern is purely spiritual and, in a peculiar way, tangential to the universal flow.

Of *my* trivium, music alone is entirely beyond language and beyond symbol.

While the condensing or "skeletonizing" of analytical evidence brings us, at its best, a limited set of symbols whipped into a semblance of structure, logic opens for us the curtain screening immense fields of the feasible. The latter is what I describe in an over-all term, as "the world of mental confines."

In the next "avenue" of the Aware Mind music does the same.

Not taking us to the pith of Being, logic as well as mathematics, and in its own but kindred sense – music, still move us closer to the mental frontiers of Being: to its norms and truth.

Spurred by an intuitional or a visionary view, I am prompted to think that the subsoil of my trivium, the orders of life, of song,

[1] Geometry, arithmetic, music, astronomy.

of ultimate knowledge, when fused, is *all there is*. Focused in one ever-present momentum, here is the only ontological reality, *the Whole*.

In an outstanding essay on logic a leading American philosopher who has shown that logic is not meaningless as to the nature of things – the late Dr. Morris R. Cohen – said something to which Kant would have gladly set his name.

"If logic were indeed only a manipulation of symbols, it would be as devoid of philosophical significance and scientific utility as chess."[1]

In the same turn of phrase, and thinking of the creative and, perhaps, a still deeper affinity between logic and music, I have a parallel statement to make.

If music were only a random shower of sounds, it would bear as little relation to True Being, be as devoid of metaphysical meaning as the clatter of plates and knives at a dinner.[2]

Logic traces the formal profile of Being for the finite, outer self. Tonal logic or musical order traces the formal profile of *All* for the aware Inner Self.

It is worth studying carefully how pressing of formal lines of a syllogism into an Euler graph – two concentric circles corresponding to the premises with the sequel manifest – repeats the order of inclusion within circumference, the essence of musical form.

Ellipsoid curve with several foci

= *Multiple syllogism, or set of reasons with conclusion (circumference)*

There is a startling counterpart to the equation symbolized in the above ellipsoid curve: again – a counterpart in thetonal realm.

1 Morris R. Cohen: *Preface to Logic* (Holt).

2 In its higher meaning no sense datum, of course, is devoid of metaphysical links, but in this "manner of speaking" there is no need for a pointing to limitations.

"Mirror inversion is a phenomenon not confined to music ... is a structural form in nature itself ... resides in the organizing powers of the mind", says a gifted and acute composer-scholar George Rochberg.[1]

Scanning a melodic twelve-tone segment from Schoenberg's *Piano Sonata* opus 25, with its mirror inversion, we find this tightly clasped and finite musical thought fully represented, in all its ebb and flow, in the above illustrated equation: ellipsoid curve of several foci – with multiple syllogism.

One thinks of still other startling affinities of philosophical vision, for which logic is major instrument, of course. I speak again of the affinity with the shaping powers of the musical mind.

Note the historical curve of Western thought through the last four centuries. The philosophy of substance or essence, or of the continuant – to put it in a newer term – was the initial pattern. Then came the principle of change or event or evolution – as dynamic antithesis to the initial rule. And functionality or organic interdependence added the perfect cyclic turn to round out the path of Western philosophy.[2]

Each of these patterns of historical thought is enclaved in a closed curve – cyclic, elliptical or ellipsoid, with two or more foci. In the tonal cosmos the equivalent is a rondo of the highest form.

Smallest unit of music's innate structure, the motive indeed forestalls its essence. The multiple life, the evolution of motive driven by tonal and modal gravitation, is music's dynamic curve. And the whole of the current shapes into a cycle.

In either, the circling line of philosophical thought or of tonal vision, the over-all pattern of the historical process embraces the functionality of motive – action – return. But of course, music is largely a function of the intuitional, and only in some special way – also, of the discursive mind.

Music's distinction and power lie mainly in the fact that its rationality is non-verbal and non-discursive.

---

[1] George Rochberg: *The Hexachord and its Relation to the 12-tone Row* (Theodore Presser Co).

[2] See a remarkably fresh *exposé* in D. W. Gotshalk's *Structure and Reality* (The Dial Press).

Seemingly "unreal" or, at least, irrational, this tonal art yet brings us closer to truth than reasoning can.

Like metaphysics busy with the ontological aspect of Being, or philosophy dealing with its cosmological and outer line, music can not yield certainties. But *it does shed inklings as to the order of the Whole.*

VI

Thus we have "ingress" into an opening in the screen of the Beyond.

Even if a not unclouded mirror of *All*, music is ideally pointless – in its absolute facet. As a term of the subliminal, the latter should mean: of absolute duration or ever-presence.

Music is the only art marked by true universality, impossible to invest with any "contents," material, descriptive, ethical, teleological . . .

As the Real itself, music *per se* has no ethical lining, no individuality, no points of temporariness, no relatedness to anything. When such element is felt in music, it is but in error: only as a response of our empirical self to the ignition that sparked music from "outside."

Here we come to understand the exact place of program in music. Even when the latter seems more than mere ignition, it debases music's true essence.

Tied to time and moment, program in music denatures it; forces a substitution of means for aim; makes an instrument that reflects things of a false world – out of something related to the true. Even as to form, program music upsets the flow of persistent return, disjoints music's innate order.

I myself have used poetic mottos and invoking narrative as title page inscriptions, in some of my symphonic works. I was careful, however, to limit the meaning of such mottos.

Thus, the composer's statement on the title page of my symphonic suite "*To a young World*" ("*A Dithyramb to Stillness*)" says:

"This work has no intention to describe or portray anything. The composer wished only to release the inner music, he heard in his subject. The poems written over individual movements are, he felt, but overtones of that inner music." [1]

It would be entirely right for me to paraphrase the Vedanta word on the nature of the universe: to say that the essence of music is related to the transcendent *All* in the same way as heat is to fire. Music, too, is "neither loving, nor cruel." I would add: "and not aware of the power of its light."

Precisely because music has no point of particularity or temporariness, a seemingly insignificant pattern, such as the opening theme in Beethoven's *Eroica* or the main motive in the finale of Mozart's *Jupiter* have become seed of an immortal form.

The "what" of all the arts in a way approaches the *Real*, that is, nears the unobservable spring of *All*. But that is not enough to invest art with the highest seership. The mode of grasping the filmy wing beyond events streaming past our "minor eye," (our finite mind), in fine, the "how" – is what really matters in music as it does in metaphysics. The instrumentality of nearing the immanent, this mainly, makes the vehicles of perception in music and metaphysics so similar; even if one is pointedly intuitive, the other – largely discursive.

Let us recall Bergson's sagacity in saying that "immediate intuition and discursive thought are one in concrete reality." [2]

It is in the very nature of music's insight and form, that more is assumed in them than can be explained. Such is also the case of metaphysics.

Music's non-verbal rationality is sourced in its freedom from *principio individuationis*. Music's soul is its order; and the seal on its body is the shadow of music's essence. Return is the visible face of the immanent *Real*. An inlet into the subliminal, music's innate order remains its only content: the immutable ontological cycle screened by illusory changes.

Even the stormiest music has over it a sheen of the apersonal and the universal: a sublimity of cosmic detachment. The pure substratum of music has nothing to do with *rerum natura*.

---

[1] Printed in the program notes when this work of mine had its première by the National Orchestral Association under Leon Barzin – in New York, at Carnegie Hall, on April 16th 1951.

[2] Henri Bergson: *Essai sur les données immédiates de la conscience*.

Through some unknown but, perhaps, discoverable roots, music does live outside and above perceptual awareness: beyond symbol and beyond language.

In the long run and each in its own way, they converge on essentials which they share: the latest wisdom of science trying to graze the Beyond, and the closest reasoning of philosophy of music.

A finite world; a void with spark-packets of energy sparsely strewn; [1] a kind of willed power of sudden shift, – a world of uncertainty, as the latest summary of science calls the observable universe.

In fine, a world spun of chance, with this correction, however, that the law of return traces – here too – the spiritual riverbed of Being. This law hems the torrent of uncertainties in tight gates of necessity.

Amidst clusters of things we either do not understand ,or cannot correlate, even the humblest intelligence yearns for a *terra firma*, hopes to have a glance of the *All*.

> *"Gestaltung, Umgestaltung,*
> *Das ew'gen Sinnes*
> *Ew'ge Unterhaltung"*
>
> "Shaping, reshaping,
> The eternal mind's
> Pastime eterne"

In rare moments of insight this yearning becomes a glow of self transcending thought. The seer power of our immediate self is then heightened to an extraordinary degree.

"Many times it has happened. Lifted out of the body into myself; becoming external to all other things and self-centered; beholding marvellous beauty, then assured of community with the loftiest order . . . acquiring identity with the divine." [2]

---

[1] One of the youngest among British mathematician-astonomers, Fred Hoyle, pictures, however, in his *Nature of the Universe*, the immense interstellar space as filled with thinly spread hydrogen, and the universe as an ever expanding but never bursting balloon.

[2] Plotinus, quoted in *History of Western Philosophy* by Bertrand Russell (Simon and Schuster).

This is the summit of insight which Santayana described less ecstatically but with wondrous precision and finality.

"In philosophy itself investigation and reasoning are only preparatory and servile parts, means to an end. They terminate in insight, or what in the noblest sense of the word may be called *θεωϱια*, – a steady contemplation of all things in their order and worth ... A philosopher who attains it is, for the moment, a poet; and a poet who turns his practised and passionate imagination on the order of all things, or on anything in the light of the whole, is for that moment, a philosopher." [1]

But sharing with mathematics and logic, with this highest plateau of investigation and reasoning, their native instrumentality in foresensing the immense field of the feasible, music reaches a still higher, even if submerged, clairvoyance of the Beyond, of its order and essence.

Through such gleams of seership we become aware of the islet of *terra firma* in our own mind, shifting and shaded. Even in our dazed separateness we can of a sudden sense the Beyond; for a blessed moment feel like a spark of it.

The nature and law of music correct the mirror held to True Being and reveal the equation of the Megacosm or Major Self: pointlessness = duration = return.

It is the revealing nature of music that leads us firmly where new mathematics and new physics (Einstein, Eddington, Milne, Heisenberg) think in guessing terms, tentatively and forever correcting themselves. [2]

Also, correcting – not repudiating – Kant!

To begin with, here is a Bergson correction: scolding Kant for a "confusing" of, or putting on a par – time with space" ... Bergson, however, prisoner of antiquated mathematics, while sharply disentangling time from space, himself confuses time, the

---

[1] George Santayana: *Three Philosophical Poets* (Charles Scribner's Sons).

[2] A brilliant young scholar of our own day Dr. Edward M. Corson of the universities of Edinburgh and New York, says in one of his books: "There exists at present a veritable jungle of literature on the theories of the so-called elementary particles and their interactions – including nuclear forces ... Apparently, what is needed (*inter alia*) is neither a *one*-particle, nor a *many*-particle theory, but rather an *all*-particle theory ..."

E. M. Corson: *Introduction to Tensors, Spinors, and Relativistic Wave-Equations* (Hafner Publishing Co., New York).

fourth dimension and a fiction-norm of perceptual evidence, with duration, substance of the *Real.*

One listens to an essentially Kantian mode of thinking when Einstein says: "The laws of nature are universal, that is, their form is invariant." [1], (i.e., not dependent on co-ordinates, axes, vectors or other geometrical paraphernalia used in fixing positions of sense data in a four dimensional continuum) ... Or when Einstein insists that in a search for the truth of the physical universe, science must start with universal concepts thoroughly illumined and correlated, and *then only* go down to observe some data in order to square the forecast of axiomatic truths with "reality."

In a similar defence of the *a priori* ways in science, Eddington speaks lucidly – and with what finality – of efficient and aware sifting of sense-data:

"Generalizations that can be reached epistemologically have a security which is denied to those that can only be reached empirically." [2]

Space can be manifold and of a variety of frames and dimensions. Time, one of the latter, can be reversed and its "direction" or curve made to depend on the line of events grouped this way or that by our subordinate, perceptual mind.

The "return" can be made a "departure." In this light, what precisely happens to the Cycle? [3] And what, then, is the meaning of this "immutable," ontological order – in its relation to time, physiological or nebular?

Thinking of variable time, with more than one dimension or a variety of "frames," one could arrive at a correct reasoning. Dr. E. A. Milne, for instance, bases *his* scale of time on the "absolute" clock existing through the vibration inside the atom; the frequency of this vibration is *tau* taken as a constant (the *tau* time). And it is easy to figure out an observer on our planet with an order of time relative to him, and the second observer – on another planet, with still another order of time related to *his*

[1] Albert Einstein: *Theory of Relativity*, Berlin.

[2] Sir Arthur Eddington: *The Philosophy of Physical Science* (Macmillan).

[3] Remember, I use "cycle" not as a strictly mathematical term but as a philosophical concept.

position, sensorium and measuring instruments. Even in the more limited world of observable physical events, time is reversible – as we saw earlier in the review of scientific evidence.[1] Again then, any moment of the time-curve can play the part of the "starting" one or of the present one.

As a key pattern of music and – more than that – as a noumenal symptom, Cycle has indeed nothing to do with the idea of recurrence: nothing to do with a symbol correlating sense-data, "reconstructing" and thus "explaining" the physical world. In the latter, however, tonal cycle is, of course, the form into which the flow of rhythmic time-segments shapes itself.

What Bergson objects to in our notion of time as duration cut into measured moments "outside" one another (intervalled time) – a notion due to intrusion of space – this Bergson objection is valid, we see, also in music's case. Valid only as we view music's *physical* flux.

The vision of a continuum spun of pure duration is, of course, upset by treating that continuum with numbers, measurements and other spatial markings. Confusing time with duration; handling the latter as the fourth dimension; fitting the tonal flow into metric and rhythmic beds and loosing sight of music's inner, of its immanent order, – these conflicts have indeed perplexed us as to music's ideal and cosmic essence.

In our highest trance of creative vision, rhythm and the rest of objective music's underpinning fade away – just as the spatial floor and vacua of the phenomenal world vanish through dimensional zero in imaginary geometry; just as the entire mirage of sense-data disappears.

Then, only the purest substance of music remains, and its contents amount to "unclocked eternity": the Absolute Present.

Like pure duration, the higher order of music, the *Return*, stays clear of the notion of time-moment as it is devoid of the symbol of space-moment.

[1] Complication which two-dimentional time might bring to our vision of the physical world, was shown also by a peculiarly encyclopedic mind living in an age of specialization, J. W. Dunn, London lawyer, authority on flying and a rather amateur philosophical scientist, in his book *The Serial Universe*.

The order of Being is not succession but duration; not consecutive change but act of relighting of the Major Self.

In music's subliminal aspect, when we shear off time and return it where it belongs – as a form of perceptual evidence and as the fourth dimension, *Return* means this only:

A re-kindling on the "screen" of our Major Mind or Ontological Self of a spark now shaded, now shining through, but always there.

In the true meaning of the *Real*, there is no such thing as recurrence. A speck of the past, or a glimmer of the future, amount only to a relighting of the Major I. This spiritual aspect of the Cycle is but a ray in our Aware Self that obliterates the symbol-weaving of our minor ego. This is when the false time-notions of the relative self are effaced.

Untangling even only the general sense-data as a web of empiric symbolization, it is entirely feasible to string the line of outer events on either a circling time-curve of consecutive moments, or a near-elliptical – of concurrent moments.

In this symbolic image of Time, circle and ellipsis are only *mental* silhouettes: circle – of straight return, ellipsis – of complex return. The ellipsis I speak of, is the thought, sequence, relational idea, not geometrical contour!

I have shown earlier that the curve of time-events, vision of the minor ego, can be likened to a circle with one point omitted: a kind of $\infty - 1$. The subliminal "time", however, should be depicted as the spirit of a *fully closed* elliptical curve.

In elliptical or ontological time the two foci of the past and the future, and the circumference – "trail" of an infinite number of moments symbolizing "the present" – all of them exist concurrently.[1]

In the subliminal aspect of "return," Being and Becoming lose difference. They, too, can be viewed as the two foci of ontological time: ellipsis merging in an all-embracing circumference of the *Whole*. One focus, the finite or perceptual self, immediate, half-awake, spinning threads of the phenomenal world, that of Becoming; the other focus, the integral and "concentric" Major Self emitting rays of the noumenal, the spark key of Being.

Using time as a provisional and working term, and making a

[1] Please, have in mind too that the term "elliptical time" means here only "co-presence." Indeed, I would not repeat Bergson's confusing time with duration. L.S.

distinction between cyclic and elliptical time, allows us to analyze the metaphysical meaning of Return; to come inspiringly close to the latter's subliminal facet.

In its ontological substratum "return" is ever here. In cyclic time – because any point on its "curve" may be and is the "beginning." In elliptical time – because the "trajectory" (circumference) is but a sum of infinite number of *presents* ever co-existing.

The spiritual essence of *Cycle,* which is the cosmic order of life, is thus revealed as Being and Becoming merged. It is right to call the blended symbol elliptic or subliminal time.

Elliptical co-existence is what duration or reality mean.

Mind's field of action, spiritual space, is just another facet of the same indefinable continuum that looms even in the cryptic boiling of the universe: in the interaction of electromagnetic, radioactive and other random packets of force. Issue of power or will, action is but a wave of the Mind, an awareness in motion.

For action is the other facet of duration, the first being – the Mind.

"Here – now" is not duration but a spatial symbol. Present, the subliminal "now," is the only duration there is; and it dwells nowhere but "*within*" the Self, as its Light. For threre is *no* "nowhere"!

How thrilling it was, long after I had conceived my *elliptical time,* to come upon the imperious word of Parmenides!

"Being is unbegotten, indestructible, whole, eternally one and infinite. With it there is no past, nor future; the whole is forever now."

Elliptical time in its subliminal aspect, the move of spirit that strings on one "curve" all the "presents" existing together, one sees again and again as the true sense of duration.

As with *quanta* upsetting the simplicity of the scheme of things in the objective world, even in the smooth – some will say, obvious – symbol of Cycle uniting the realm of music, and, for the reasons here presented, foresensing the Beyond, – there are dark points confusing the order that transpires in music's flow.

One finds, of course, adjustments in the philosophy of music similar to those reconciling *quanta* with relativity: a moody medium with a continuous one. Elliptical or other curved tonal waves are approaches to the *Ultimate* just as elliptical time is.

And that ultimate concept makes us understand and place illusory discrepancies.

Broadly speaking, an inspired thought of the real sub-stratum of *All*, life or duration, or Ever-Present, seems to be the only *absolute* in relation to which "uncertainty" makes sense. So is a ray: for a moment screened by a passing shadow, yet always there.

In music's primordial unit, the motive, one observes the power or awareness in motion, that shapes the motive's destiny: its creative trajectory, as it were.

Through this life-power the motive first multiplies itself and grows, then shapes the whole into an inescapable form: at its highest – into the sonata pattern with its full circle of dialectic unfolding of motives and their aggregates. This order repeats the pattern followed by the all-embracing, infinite Inner Mind. It repeats the latter's wave, predestines the route of Being.

Not only discrepancies, ever polarities are dissolved in a foresense of the subliminal.

Idea and life-strain; mind and action; essence and form; still-standing and movement; propelling force and the route of thought – all converge on one focal point. Elipsis dissolved circle. Oneness dissolves duality.

The subliminal equation of Being = Becoming = Being is what the basic order of music adds up to. And this happens to be also the essence of the sonata. Only in music does the inner life-whirl of the motive predestine the order of the whole *inexorably* – exactly as the life-whirl of cosmic awareness shapes the birth of a child or the path of a planet.

Only in music! When in his lucent image Pushkin traces a rondo of the second form – a purely musical order unusual for poetry – it is a wave of musical mind that spurred him, not any other.

> Sing not, fair, heedless child,
> The melancholy Georgian chants of yore,
> They revive, recall a hapless tide
> Of life remote, a long forgotten shore.

Your singing cruel, laden
Rallies and restores far steppes,
Night swept with a silver light,
And the traits of a distant maiden.
Sing not, fair, heedless child,
The melancholy Georgian chants of yore.[1]

This, as I said, is a strictly musical form. Audibly sisters, music and poetry, have, however, nothing basic in common, save faint aesthetical roots. The native rhythm, route of flowing, mould of sonority, formal silhouette – in an innately musical creation, are so very distant from those of poetry!

Sing not, fair, heedless child,
The melancholy Georgian chants of yore ...

This is Pushkin listening – not to an echo of our humble lower ego, nor, indeed, to a fair child's song; but – on a high wave of self transcending awareness – to the whirr of the subliminal, to the *Mind of All* mantling Cosmos.

And Beethoven intent upon the voices within himself, hearing the murmur of Creation, and singing of its dark ebb – in the opening bars of the *Ninth Symphony*!

And Plotinus harkening in full abandon to the voice of the Ultimate!

What does one read in the eternal pattern of music? – To become, yet to be; to move, yet to stand still; to change, yet to remain the same thing. In fine, the meaning of substantive existence.

. . . .

"The circle which Giotto drew with a sweep of his hand was not marvellous because it was almost mathematically *correct.* The wonder of the performance was typified by that diminutive *divergence* from correct formality which was the subtlest gift of Giotto's genius, and expressed his delicate personality. In just the same way, the disk of a sunflower is not a perfect circle because it is a portion of a living plant." [2]

Yes! ... For the spiral, an unfinished, unclosed curve and

[1] Lillian Saminsky: *Poems and Adaptations* (Oxonian Press, Oxford, England 1931; memorial edition, New York, 1947).
[2] Sir Theodore Andrea Cook: *The Curves of Life.*

seemingly primordial type of form in nature and art, is only a swing of the unaware mind, a move darkly yearning for finality.

In this set of sense-data one reads something of the deepest meaning. One watches a strain never to achieve an ideal and finished turn, circling or elliptical; never to turn into rays of the Ultimate.

With this in mind, let us set our eye, very closely this time, at the structural distinction between the *"tonal sisters."* In spite of some fragmentary and faint similarities of formal profile – the golden mean ratio, for example, linking a Dante poem with a sonata pattern in one of Beethoven's quartets –, the golden section is as much of secondary, of phenomenal point as any generalization of sense-data is: like the logarythmic spiral of growing plants or the equiangular curve of the *nautilus pompilius* sea shell.

On the contrary, the implacable cyclic urge in musical unfolding is unknown to poetic form as such. *This urge is of primary and transcendental point.*

While the formal lines in poetry and plastic arts remain symbols of phenomenal evidence – even when they casually intimate a musical outline, only the strictly cyclic order native to music has the glow of subliminal testimony.

Touched by phenomenal relativity, the "curves of life" in nature or art do not even graze the Beyond. They give us inklings only as to the direction of the outer flux of Being, not its inner order.

While the logarythmic spiral, a non-closed curve, implies a direction of life and growth, circumference (circle or its mode, the ellipsis) breathe the order that symbolizes Immanence, or Return, or Relighting of the Major Self.

In the phenomenal world nothing is "closed", and there is no absolute simultaneity.

In the subliminal world everything is closed. Nothing exists there but absloute simultaneity. Which is but elliptic time – a circumference of presents.

Perception of "happenings" and ensuing reasoning hide inferences that are really but statements of belief.

Isolating categorial traits in our perceptual, that is, immediate, experience – traits held to be of essential reality – is not outside theoretical science.

My description of nature's telling phenomena, essential reality of music's heart and form, music's cyclic and elliptic stances, shed light, I am proud to say, on the same domaine of the categorial as Whitehead's thought did, and for the same purpose.

True, my evidence is not of absolute reality, and my argument may not be rockbound. The first is an aggregate of affinities, and the second is bound to a file of potentials.

Yet in all its segments, symptoms, links, variations, the rule of the Ring intimates the nature of our Higher Self. Our ultimate awareness comes into full play, and we then sense the Infinite I as the immediately given. Latent or lighted, *this* is the Only Existent.

It is music that causes our I "to regain oneself and to get anew into pure duration." [1] Of course, Bergson did not speak of music, did not even think of it.

The basic order of music not only equates that of Being; not only reflects the play of the all-aware Major Self with the worlds observable and subliminal – dissolving the former in the latter. In turns, the supreme order of music screens and relights the *Ever-Present*.

If philosophy serves and affects the aware intellect, music serves and sparks the innermost mind.

Like in metaphysics, it is given to music to foresense the salient, the ultimate realities of Being.

Not unlike science, plastic arts would only approximate the data of perception and relate them in an acceptable concept. Music lifts its glance up to the very mirror of Being.

---

[1] Henri Bergson: *Essai sur les données immédiates de la conscience* (Paris, 1889).

The Roots of Arithmetic

In a strict system of arithmetic the natural series of numbers is taken for granted; the terms of equality and non-equality are set up as well as the idea of direct and reversed operation.

Where does the idea of prime numbers and of the natural series itself come from? What is the nature of the setting up of terms? And the philosophical meaning of the broadening of the number concept? How far and in what respect do the foundations of a strict arithmetical system warrant its certitude? All these questions belong essentially to epistemology. To avoid misunderstandings resulting from a faulty or not fully explored viewpoint, such as is sometimes held by partisans of philosophical empiricism, and more often still by philosophizing mathematicians, it is necessary to note the following.

The questions about the foundations of arithmetic reduce themselves to two which probe two different psychic media and should not be confused: the actual origin of the number concept, and the nature of arithmetical judgment.

The first is an anthropological problem, that is, a scientific and a relative one. Its solution need not necessarily influence the answer to the second question, which is a purely philosophical one.

The first problem is to be solved by the facts of experimental psychology, anthropology and history of culture; it has to do with the number symbol, or general idea of number: with half-sensual images of countable groups or the material of number concept.

The philosophical problem is to be solved by probing the logically and mathematically treated idea of number; it concerns the product of reflection over the material of number concepts.

To separate the two tasks and their objects is not only a question of method but also, a means of stabilizing our viewpoint, of throwing light on the details of our root problem, and of revealing at times the misunderstandings and the mistakes of both, the protagonists and the opponents of the empirical theory of the origin of number.

By this I do not mean to imply an insistence upon a complete and basic independence of our key problems, but merely an affirmation of their *possible* autonomy.

In other words, the question as to the nature of arithmetical judgment *can* be answered outside the data or the origin of the number concept. But of course, the latter inquiry has an important gnosticological value even if arithmetical judgment bears the stamp of empiric origin.

II

By indicating the two types of arithmetical formations, the number image and the number concept, I did not wish to point to the path of historical development: the flow of arithmetical concepts has possibly expressed itself not in the above forms alone.

There may be transitory forms.

The forms described above present the pure source of arithmetical propositions and link two aggregates of qualities of different character, neither of which can be deduced from the other. And this, in spite of the fact that the number image is necessarily at the bottom of a strict system of arithmetical concepts.

Theoreticians feel this necessity, and they lay at the basis of their systems symbols that serve as a sort of experimental soil for number concepts. They try thus to put through the empiristic viewpoint as to the roots of mathematics.

As material for arithmetical concepts Helmholtz presents the idea of "quantity of objects" (*Anzahl der Dingen*) [1]; Stolz – the image of "many" (*Vielheit*).[2]

But these fundamentals are not free enough from reflective

[1] Helmholt "Anzahl der Dinge", *Wissenschaftliche Abhandlungen*, III. Band.
[2] O. Stolz: *Vorlesungen über Allgemeine Arithmetik*.

elements – to present the pure material of number concepts; and they do not let us into the hidden source of arithmetic far enough to explain the nature of its propositions.

The number image, its link with the number concept and their very existence as different mental media, are best seen at their birthplace, in the mind of a child or a savage.

Anthropology knows savages who have an idea of a few small numbers only: three, four, five.[1] Can we be sure that because of this those savages can not imagine larger numbers?

The inhabitants of the Tonga Islands know up to a hundred thousand. Does this mean that they have a true number concept of such a quantity?

These queries can be answered through the above delimitation.

The number image is the sense-picture of a group of objects almost unconnected with the concept of the division of such a group into distinctly separate objects which thus become units. The number concept does, however, carry with it the awareness that a number consists of units or entities. This knowledge has been borne by reflection which disintegrated the primal number group or number image.

When a savage of the lowest cultural strata, who could name no number above "three" or "four", speaks of larger groups as "many," he does have an image of numbers greater than "three," but his feeble reflective mind cannot divide a larger number into "ones." Nor can his feeble abstract memory retain the course of division necessary to form the number concept.

In its concrete and vivid form here described, the number image or symbol, is of course, the material of arithmetical concepts; is what the conscious mind turns over. This material is not primitive, however.

If one frees the number image from its sensual, outer appearance, from symbols pale yet not devoid of weak material contents and making up the countable group, there remains a certain mental residue that hides the true root of the number concept.

---

[1] Taylor: *Primitive Culture*, vol. I, chapter VII.

This residue could be described as the impression of co-existence of distinct images in a group, or to put it more precisely, *the pure impression of divisible co-existence.*

The defining of the nature of this primal material of the number concept has a bearing on the question of the number's physical existence.

Mill insists that the number is an abstract of sense-data: material things contain something which suggests the idea of number.[1] Psychologists of the Kantian school reply that number is a product of reflection by our awareness on our own action: its material can not therefore exist physically.[2]

These opinions do not exclude each other, for they pertain to different things: the former, to the number image; the latter, to the number idea. The opponents are not aware that they talk of things belonging to different mental orders.

The number symbol (or image) depends on the relative position of objects in a group that can be counted. Mill is perfectly right in basing his reasoning on the point that we can arrange three pebbles in a way which makes for a description of the group as "two pebbles and one pebble." And one can arrange them so that they present "three pebbles." This example shows clearly that the number image is formed from outer experience.

But it is important to know definitely which aspect of outer experience is presented in the number image. Let us take the following example.

Let us lay ten cannon shells close to each other or heap them together into a pyramidal or global mass. This group would leave the impression of "one" not "ten." i.e. of *one* heap. But when the shells are placed pretty close to each other yet clearly separated, the group would not fail to leave the impression of "*ten*".

It is clear, then, that at the bottom of the number image or number symbol lies the pure impression of divisible co-existence, and that it expresses solely the character of co-existence of things in a group. But it is equally evident that as an abstract from impressions of divisible co-existence, the number does not exist physically. Such impressions are totally without contents called "outer experience."

[1] John Stuart Mill: *The System of Logic*, vol. I, chapter vi.

[2] Prof. Tchelpǎnov: "The a priori elements of knowledge," *Problems of Philosophy*, Vol. IV (Moscow).

One cannot say that number is a quality of things, or that it is not: neither makes sense. Number is a quality of experience as a whole, not in parts; for number's very nature is in that it states the division of experience into objects and the uniting of things in experience.

This is why Mill is wrong in affirming, and the psychologists in denying, that number is an abstraction of something that exists physically, and therefore exists physically itself.

The empiricists make a correct statement of fact – the origin of number, but draw the wrong conclusion. The psychologists, establish a wrong fact but draw the right conclusion. Both parties are at fault in the way they link fact and consequence.

However, no matter what the material of the number image is, it is still given in the physical relation of objects in a countable group.

The nature of this physical relation reveals the conditions of countability. A correct definition of them is to be found in this precise statement of Helmholtz:

"The objects of counting should not fade, nor merge, nor divide themselves into fractions. New objects should not appear, so that a clearly separate one should correspond to each countable thing labelled by a Greek letter." [1]

Which means that terms of countability are those of division in co-existence.

What is remarkable and telling, is this. Evoking the number image we fancy a group of sticks, or points, in fine, of objects whose material contents at once vanishes from our consciousness leaving only the pure impression of divisible co-existence. The latter is of material contents so filmy that its idea consists of only formal traits expressing the law of co-existence of things, the law of their pure relationship.

If such a quality does not pertain to the nature of the things themselves, but of the outer world (the sense-data *as a whole*), then the above law of co-existence must be a pre-condition of the outer experience.

---

[1] „*Sie* (*Objecte dez Zählens*) *dürfen nicht verschwinden, oder mit einander verschmelzen; es darf keines sich in zwei Theilen, kein neues hinzukommen, so dass jedem in Form eines griechischen Buchstaben gegebenen Namen ausdauernd ein und nur ein abgegrenztes und als einzelnes erkennbar bleibendes Object entspricht.*"

Helmholtz: *Zählen und Messen*

III The peculiar nature of the number concept lies in its clear idea of' 'one" and in the awareness that the number consists of ones. Could the empirical material that feeds the number symbols or images, or the symbols themselves be the source of the above mentioned peculiarity of the number concept?

To begin with, it is easy to show that concrete ambiance or outer experience are a necessary but insufficient condition for the forming of the number concept.

All savages have five fingers on each hand, ten on both. Yet some Australian natives cannot count five, or even four; some of the most primitive tribes of Brazil have no names for numbers higher than "three" and say: "many." Does this mean that such savages have no more than three fingers on a hand; or that in their life and sense-experience there were not sufficient incentives for the forming and discerning of numbers higher than "three"?

Had concrete experience, such as counting on fingers, given us the number concept with its key characteristic of division into units, then any savage would possess the knowledge of numbers up to ten, at least: the more so, that necessity for segregating and naming the different numbers is called for by any kind of sense -experience. Facts stating the reverse of such a situation show that, firstly, in the number image (or symbol) there is no clear awareness of "one"; secondly, fingers serve only to fix and express the already existing awareness of number, not to acquire such knowledge.

In spite of the fact that the awareness of number cannot apparently be born from mere outer experience, one could fancy an arithmetic devoid of the idea of number, and consequently – devoid of the concept of "one." It is *à propos* to mention at this point that "one" is a purely negative concept, for it merely states the absence of divisible coexistence.

Such arithmetic should be called "physical" as its concepts and actions would have a purely sensual character: would operate with number images. In such arithmetic addition would be a tossing together of two or several groups and observing the resulting image. Subtraction would consist of separating one physical group from another, and a contemplation of the remainder. Multiplication would amount to repeated massing of groups; division – to a repeated separation.

True, in subtraction of this kind, before one segregates the group to be subtracted, one has to assemble it and have a preliminary notion of "one." But in this case, the latter image would be vague, indistinct, almost devoid of the traits of a countable thing. Besides, if subtraction is replaced by a groping for the addendum making up, together with the subtracted group, the initial number, in such an action the idea of "one" would not be at all necessary.

This is how arithmetic would look, were its basic empirical material handled without the aid of full awareness. Theoreticians sense this, yet with a surprising ease they flit from number symbol to number concept.

Stolz explains the formation of number concept very simply. "Any of such "manies" or "quantities" (denoted number) has one only counterpart in the natural series. This one finds by drawing out, one after another, units assembled in the quantity and marking each unit with the symbol 1." [1]) But in the concept of "*Vielheit,*" i.e. plurality or "quantity," the trait of divisibility through reduction to "ones" is not inherent; there is no awareness that this trait is formal, that is, not dependent on the concrete nature of the things forming a group.

In the number concept divisibility into units and awareness of the latter's formal nature *are* inherent.

Such traits cannot possibly develop through a simple transition from number-symbol to number-concept.

These qualities come into being owing to the action of consciousness dissolving number symbols generally devoid of awareness of the composition of number.

Only when the reflecting consciousness notes and counts its halts, it notices also that the number is composed of "ones," and in the same way as the group of objects expressed in the number is made up of single objects.

The very idea of natural series is also derived through reflection. This series' formation would not be feasible as long as consciousness lacks the clear idea of unit and of the realizing how numbers consist of units.

---

[1] „*Jeder solchen Vielheit (benanntenZahl) entspricht nähmlich eine ihr gleiche natürliche Zahl, welche gefunden wird, indem man von den der Vielheit angehörigen Einheiten eine nach den andern ausgegreift, sie mit dem Striche 1 abbildet und hierauf bei Seite legt.*" Otto Stolz: *Vorlesungen über allgemeine Arithmetik,* Band I.

Which explains why primitive men with more fingers than three or four about them and an environment that prompts the creating of ideas of larger numbers, often know no higher one than "three"; discern none of the numbers larger than that, calling them all "many."

It is quite possible that they can hold in their imagination the image of many numbers but, as I have already said, their feeble power of reflection and of memorizing does not enable them to dismember the image subjected to a counting or remember the number of halts of awareness. This is why they cannot gather that, for example, number "four" or more is related to four or more fingers; that a number is composed of units just as an ensemble of fingers in made up of single ones.

For the same reasons the mind of primitive man lacks the concept of natural series and does not see the difference in numbers beyond a certain limit.

IV

Number formations at the root of our arithmetical concepts have been so far looked into only genetically: as a source of these concepts.

To follow their logico-mathematical work-out is to sound the elements of the system of arithmetic. But such a search must be preceded by some thoughts on the meaning of the broadened concept of number to the problem of the roots of arithmetic.

Studying the foundation of arithmetic, psychologists of old usually concerned themselves with whole and positive numbers only. They pointed to the fact that other number concepts originate in the whole and positive number.

Even if such feeble remark could be supported by the algebraic number's being a result of a finite file of operations with whole and positive numbers, the fact remains, firstly, that transcendent numbers lack the latter trait,[1] and secondly, that the above statement totally evades the question as to the meaning of broadening of the number concept.

This question is, however, very important.

If the arithmetic of fractional, negative, irrational and

[1] Volkoff: *Evolution of the Idea of Number.*

imaginary numbers has neither correlatives in outer experience,[1] nor natural applicability to experience, yet is marked by the same seal of certainty as arithmetic of the whole, positive number; then one is bound to conclude that, in general, arithmetic is a closed science: a purely mental construction whose partial applicability to experience is casual.

It is also asked: if fractional, irrational and other such numbers have no roots in sense data, as whole, positive numbers have (with the sensations of divisible co-existence at source); that is, if fractional and irrational quantities are not really numbers in a basic sense, then what meaning can be given to operations with such numbers?

This problem – just as the corresponding query concerning the foundations of geometry – is muddled by the ever present error of philosophizing mathematicians: taking the broadening of a mathematical concept for a deepening of a philosophical one.

To those mathematicians – only the highest abstraction of the number concept seems important; and in the definition of number and system of arithmetic, all ideas of number floating in science are deemed to be of the same philosophic potency.

To refute this I must point out that, on the one hand, the result of generalization for all concrete number concepts is the idea of number function,[2] not number by itself. On the other hand, at the bottom of our idea of number, therefore also – of the nature of arithmetic, lie specific traits, not generic ones: the very thing that tells the concrete type from the abstract general. Also, it is to be particularly noticed that the number concept

[1] It is possible, of course, to select an aggregate of sensations corresponding to a given number, as analytical geometry does building the curves of any function. There are, however, no sensations that would manifestly be psychological roots of irrational or imaginary numbers.

[2] In the process of generalization the number concept loses its basic meaning, and the mathematicians find it necessary to resort to the idea of number function. This can be seen from the traits of the *Zahlkörper* indicated by Dedekind (*Stetigkeit und Irrationale Zahlen*, § 1), and from his passage in the same chapter where – with difficulty – he tries to solidify the operations with concrete numbers: "*Die abschreckende Schwerlichkeit aber, welche dem Ausspruche eines solchen Satzes anklebt, überzeugt uns, das hier etwas geschehen muss, um der Sprache zu Hilfe zu kommen; dies wird in der Tat auf die vollkommenste Weise erreicht, wenn mann die Begriffe der veränderlichen Grössen einführt.*"

"The dreadful heaviness that marks the expression of such proposition convinces us that something must be done to help the wording. (Only the wording? L.S.). This can best be achieved by an introducing of the idea of variable quantities." (*Ibidem*).

embraces whole, positive numbers and its contents lie mainly in the trait of "discreteness," that described as "*Einsetzung*" and "*Abgeschlossenheit.*" [1]

This is where our right lies to consider adequate – in the problem of the foundations of arithmetic – an analysis of the whole, positive number only.

To prove this proposition, one has to scrutinize the number in its rôle of an arithmetical function.

Using the favorite proceeding of theoreticians in introduction of analysis, we are going to glance at the comparative nature of the number system and of the straight line.

Dedekind affirms the analogy of these two functions and deduces it from the following points:

I. a) If a > b, and b > c, then a > c

b) If point p lies to the right from q, and q again to the right from r, then p also is to the right from r. And one says, in this case, that q lies between points p and r.

II. a) If a and c are two different numbers, then there exists always an endless row of numbers quantatively placed between a and c.

b) If p and r are two different points, then there exists any number of points lying between p and r.[2]

Dedekind's third pair of statements of this kind is really contained in the first.

From this table one gathers, however, that there is no complete analogy between the numbers' system and the straight line.

Where such analogy looms, it is purely symbolic, and a convention.

But number is a psychological fact rooting in our sense experience. To deduce its nature from a comparison with a conditional and rationalized description of another psychological fact (straight line) seems deliberate.

In the above point I the analogy of the two propositions lies merely in this. A segment of a straight line may be represented by a number unit, and the direction from the point selected as the line's beginning – by a symbol. As to point II, not only does one

---

[1] Dedekind: *Stetigkeit und Irrationale Zahlen.*
[2] Dedekind. *Ibidem.*

find no analogy between its two propositions; such analogy would have contradicted the nature of number.

The proposition that if numbers *a* and *b* are not equal, one can interpolate any number of intermediate quantities, this proposition does not in the least conflict with the trait of "discreetness"

As a matter of fact, taking two closest irrational numbers which because of the infinitesimal difference in their quantative value admit an infinite proximity, this closeness is yet limited by the ultimate estimate of these two numbers. The difference of their ultimate limits cannot be zero, as it would mean that the square roots of differing numbers could be equal.

A continuity in a system of numbers is not obtainable: for the interpolating of *all* possible number values between the infinitessimally close numbers, is unthinkable. But interpolation of all feasible segments between two points is really obtainable: these segments *exist* in the field of vision or imagination – even if we cannot locate them.

In its very nature the number is "discrete" just because its roots lie in sensations discontinous, neatly separable from other sensations.

Analogy between a numbers' system and a straight line is possible only because both could be covered by one mathematical generalization: the function of one variable.

But owing to his characteristic way of thinking, the mathematician mistakes a mathematical abstraction for a logical. Thus Dedekind proves – in spite of the nature of number – that a numbers' system is continuous. True, he correctly carries over the traits of the system of rational numbers to the irrational group having shown that the latter presents qualities of the function of one variable.[1]

It is clear then, that the typical character of number is expressed in the character of the whole, positive number with its trait of discretion and with sense-roots manifest – no matter what the aspect of outer experience expressed in such number is.

If, however, numbers fractional, negative, imaginary and so on, are not numbers in the basic sense, and are but arithmetical functions pointing to a quantitative connection not translatable into a number from the natural series, then operations with them

[1] Dedekind: *Stetigkeit und irrationale Zahlen*, § 5.

and the result of such operations have a purely conditional meaning.

The proposition $4 > 3$ means that number 4 dwells on a higher place of the natural series than number 3. This is not conveyed by the proposition $-3 > -4$. The latter means only that the result of subtraction from any number *a* of one larger than *a* by "three," would always be larger than the difference between the same *a* and the number $a + 4$ — *if* two numbers embodying these differences existed.

The above discussion makes us deem the analysis of arithmetic of whole, positive numbers *philosophically sufficient.*

V

There is still a rather slight difficulty – in the question as to precisely which system of arithmetic must be looked into.

While the totality of arithmetical truths is invariant, the concepts at its basis are so multiple that the logical ensemble of arithmetics rooted in those concepts may take on diverse light and aspect.

Thus, it would be possible to consider the systems built on concrete concepts by Helmholtz and Stolz and unfolded in their works quoted earlier in this essay. Or it would be possible to study the remarkable ensemble of concepts taught by Dedekind and carried by him to the confines of abstraction; [1] or even, the system of mathematical logic – on the premiss, that arithmetic is but an interpretation of the latter.

In all these systems, however, one finds a certain group of elements common to each, that is, sufficient and necessary for the formation of any such systems. These elements comprise basic terms and propositions. Arithmetic's axioms are ingrained in some of these propositions.

Of the five axioms defined by Helmholtz, I, IV and V are issues of the term of equality; II and III are basic propositions.

Only the equality concept can be deemed a prime term not reduceable to component parts. It is easy to see that other terms are but derivatives from that of equality (the term of inequality,

[1] Dedekind: *Was sind und sollen die Zahlen.*

for example); or they are combinations of the former with secondary and subsidiary ones, such as the term of operation.

As a rule, theoreticians set up the term of equality uniformly.

Termed "equal" is what could be expressed in the same way or can replace each other in a proposition" — says Grassmann.[1]

Almost identical is Stolz's definition:

"Thing *a* is identical with thing *b* if all that is thought of *a* pertains to *b*, and all that pertains to *b*, is thought of *a*." [2]

"The symbol of identity a = b in the study of the number must signify only that "*a* is the same number as *b*" - says Helmholtz ("*Das Gleichzeichen a = b soll in der reinen Zahlenlehre nur bezeichnen*: "*a is dieselbe Zahl, wie b.*" [3])

But both kinds of theoreticians, those who would tie the basis of their systems to the soil of outer experience (Stolz and particularly, Helmholtz), as well as the others - who build their arithmetic on a strict logical foundation of abstract concepts (Dedekind), make the same basic mistake. Their error is a result of dim awareness of their own viewpoint and of a mistaking of mathematical terms for principles of logic.

In the above definitions looms the thought that the idea of equality (*Gleichheit*) is a concrete concept, that is, one pertaining to the material contents of the objects that are being compared. This is how this term is mistaken for that of identity (*Identität*). And wrongly so, if only for the reason that the polar twin concepts of nonequality and of difference coincide neither in scope, nor in contents.

To begin with, the idea of identity is purely negative. This can be seen from the correct definition by Siegwart:

"The symbol of identity implies not only that the same object existed for a longer time or repeated itself. "Identity" originates first through the negation of difference in contents of two or more images following each other in time. It is in such action that the identity concept begets its meaning." [4]

---

[1] „*Gleich ist dasjenige, von dem man stets dasselbe aussagen kann, oder allgemeiner, was in jedem Urtheile sich gegenseitig substituirt werden kann.*"

[2] „*Ein Ding* a *is dasselbe wie* b (*identisch mit* b) *und* b *dasselbe wie* a, *wenn alles, was von* a *gedacht werden kann, auch von* b, *und wenn alles, was von* b *gilt, auch von* a *gedacht werden kann.*"

[3] Helmholtz: *Zählen und Messen.*

[4] „*Die Vorstellung der Identität setzt nicht bloss voraus, dass dasselbe Object längere Zeit oder wiederholt gegenwärtig war, sondern sie entsteht erst durch Negation des inhalt-*

It occurs to us, however, that in the arithmetical proposition $5 = 3 + 2$ there is no such negative action. For this formula *does not deny difference of contents* in the equated symbols of 5 and $3 + 2$. It only affirms the fact the two equated figures are composed of units in exactly the same way. If therefore one can not insist that 5 is *the same* (*dasselbe*) as $3 + 2$, it would follow from the above defining of the term of equality that "5 is not equal (*nicht gleich*) to $3 + 2$."

But all such considerations are partly arguments *ad hominem*. The following is essentially clear.

The proposition $5 = 3 + 2$ is not an identity for the reason, to start with, that every identity is an analytical judgment. But the concept $3 + 2$ by itself by no means contains the trait of "five" (as Kant had shown in his well known thoughts on the matter). Should we, however, think that this proposition has for its ground (*Grund des Urtheils*) the general image or concept of 5 (*Gesammtanschaung*),[1] it is clear that the above proposition cannot be a statement of identity: the latter is self containing.

An important conclusion suggests itself.

As long as the term of equality does not touch the contents of the image bound in the proposition of equality, this term has apparently no roots in the material of sense data, only in their form. It follows that the mathematical concept of equality is a purely formal one; it applies only to the trait of countability.

Thus one could say that $3 = 3$ or that 3 apples are identical with 3 apples, but it would be odd to state that 3 apples are equal to 3 apples.

Giving closer thought to the last example, one perceives that only in case when the three given apples are identical with the other three in every detail of size, form, color and taste, we would feel compelled to say that these 3 apples are equal to those 3.

An empiricist would seize upon this and say at once that, of course, the equality concept is material and pertains to contents of sense data.

---

lichen Unterschiedes zweier oder mehrerer, zeitlich aufeinander folgender Vorstellungen und hat ihren Inhalt in dieser Tätigkeit." Siegwart: *Logik*, Band I, § 63, 3.

[1] This is a judgment where, in Siegwart's words, "the subject and the predicate with their relationship are but components of the given complex image." (*Dessen Subject und Prädicat mit ihrer Beziehung nur die Bestandtheile einer gegebenen complexen Vorstellung darstellen*).

The example proves just the reverse.

As a matter of fact, we have affirmed the equality of the two groups of apples the moment their details appeared completely identical. But then they have turned into merely number images or countable formations since the identity of the apples' qualities made vanish from our mind the awareness of their material contents. Only their formal trait, the manner of their coexistence and countability remained with us. It is the same mental process by which, wishing to imagine a countable group, we evoke in our mind a series of points, identical small strokes, and so on.

It is clear, then, that the proposition of equality expressing only the fact that the manner of co-existence of things in the two groups, is the same – this proposition has merely formal contents: it compares objects as to traits that are formal by nature.

VI

The other elements of a system of arithmetic are the basic propositions.

No matter what their contents are, they are judgments as to form. It is important to trace their link with the general theory of judgment and with the concepts of traditional logic.

But risking a reproach from Kantian old-believers, I must say that I cannot see the autonomous existence of synthetic judgments, be they *a priori* or *a posteriori*.

If, in an *a priori* proposition a point in the predicate visibly leaves the bounds of the subject, such a judgment is but the conclusion of a hidden syllogism where the premisses, being self-evident truths, vanish from the field of awareness. These very premisses pre-ordain the subject of a synthetic *a priori* judgment.

As to propositions synthetic *a posteriori*, they too speak only of the contents of their subject, in which the defining trait is merely omitted. That is, the general trait is left out when the judgment is general; and the private one when it is limited.

In these *a posteriori* propositions their synthetic nature is but deceptive; for the trait not seemingly belonging to the subject of the judgment was already tied to that subject by apperception, that is, prior to the act of judgment.

Then also, these judgments are ruled by the principle of contradiction. The negation of a synthetic *a posteriori* proposition contradicts the particular experience if these judgments are private, and to the total experience if they are general.

Such a statement may seem odd to those thinking with Kant that "propositions derived from experience are always synthetic; it would be absurd to base an *analytical* judgment on experience: for such judgment does not have to step beyond its subject." [1]

My above statement is, however, correct. For the synthetic link of the predicate's traits with the subject was already given in the perception; the judgment only bares the contents of sense data.

There is still another, a simple and telling consideration. If the predicate of an *a posteriori* proposition originated in experience, it has been included in the subject prior to judgment. The latter is thus analytical. If this predicate did not come from experience, then again, the judgment is not possible as long as the trait of the predicate is not already contained in the subject-concept.

Kant says that the judgment "body has weight" is synthetic *a priori*[2]. Confronting this with the proposition "geometrical body has no weight," is easy to see that in the first proposition the defining trait "this" (body) or "physical" (body) has been omitted. When restored, that makes the judgment analytical: private ("*this* body has weight"), or general ("*physical* body has weight.")

Thus any such judgment is analytical. From my standpoint, it should be called *a priori* or a *posteriori* dependent only on the nature of the synthetic connection between the traits of the subject and the predicate.

If such connection is ingrained in reflection that formed the general concept, subject of the proposition, the said judgment is *a priori*; if the connection came from perception that formed the image-subject, the judgment is *a posteriori*.

These remarks are presumed to be fenced off by Kant's words that within the limits of experience there are no strict definitions;

---

[1] *„Erfahrungsurtheile sind insgesammt syntetisch. Denn es wäre ungereimt, ein analytisches Urteil auf Erfahrung zu gründen, weil ich aus meinen Begriffe gar nicht herausgehen darf um das Urtheil abzufassen."*

Kant: *Kritik der Reinen Vernunft*, Einführung, IV.

[2] Kant: *Ibidem*.

that for this reason the concept-subject of a judgment derived from experience has only a nominal definition sufficient for telling it from another.

Influenced by Kant's remark, Schleiermacher finds the difference between analytical and synthetic judgments relative or rather, passing. For example, the proposition "ice melts" will be one or the other depending on whether the concept of ice includes the idea of temperature-conditions of its existence or not.

But one finds here the same mistake. When this judgment is private, it will read: "melting ice melts"; if general, it will say: "ice can melt." In both cases the predicate was included in the subject-concept prior to judgment: in the first case by perception; in the second by reflection.

Therefore, I must add, subconsciously (by preceding perception or reflection) a trait is included in the subject, and namely, one to become the predicate of the proposition. And this in any case: whether the induction forming the subject-concept was already realized or not.

The above considerations result in two leading thoughts.

1) If the basic propositions of arithmetic are analytical, then, be they result of outer experience or *a priori*, their source lies in the subject of the proposition, that is, in the number – itself rooted in the impressions of coexistence.

2) If the basic propositions of arithmetic are synthetic *a priori*, i.e. they are conclusions of syllogisms with self-evident premisses that vanished from consciousness, then the source of the conclusions is that of the propositions. When the premisses are "located," the source lies in the subject of the minor premiss. This premiss is analytical in any case.

Let us apply these findings to the question of the source of the basic truths. They amount approximately to five axioms.

I) Two quantities equal to a third are equal.

II) The associative law of addition: $(a + b) + c = a + (b + c)$.

III) The commutative law of addition: $a + b = b + a$.

IV) Equals added to equals result in equals.

V. Equals added to unequals result in unequals.[1]

---

[1] According to Grassman, axioms II and III amount to $(a + b) + 1 = a + (b + 1)$ Helmholtz advances two more axioms: (*cont'd on p. 72*)

α) Helmholtz shows that the Ist axiom is a corrollary to the equality formula set up by him and other theoreticians. The proposition $\dfrac{a = b;\, b = c}{a = c}$ says essentially this. If *a* is the same as *b*, and *b* the same as *c*, then *a* is the same as *c*.

But if such be the case, this axiom would be sheer verbiage, sheer tautology, not truth enlarging our knowledge.

Helmholtz's error, as I have shown earlier, consists in his mistaking the empirical concept of identity for the mathematical concept of equality.

Axiom I expresses conditions for the comparing of objects in their purely formal qualities; namely, the certainty that the same groups of varying objects are composed of their units in the same way as countable objects are. This proposition is formal and synthetic, for its predicate goes beyond the subject.

According to the results of our discussion, there must be hidden premisses to such a proposition. In fact, the latter is the result of a syllogism where the major premiss is the axiom of independence of numbers (i.e., countable groups) from the quality of the objects being counted; the minor premiss defines the terms of countability (see p. 59 and presses the objects being compared into the concept of countable groups.

Which means that the source of axiom I is a self-evident premiss — the axiom of independence.

β) Propositions presented in axioms II and III contain a derivative term of sum to which Helmholtz gives the meaning of order. For example, the sum $a + b$ is the bth number of the natural series counted beginning with the number $a + 1$.

But the axioms mentionned include again some synthetic proposition going beyond such idea of sum. If we take several pairs composed of different objects but ranged in similar groups – say, 3 pebbles and 2 pebbles, 3 apples and 2 apples – we find the

---

1) "If two numbers are different, one has to be found higher in the natural series than the other."

2) "If number *c* stands higher than *a*, then one can figure *c* as the sum of *a* and some whole, positive *b* still to be defined" (*Zählen und Messen*).

Both propositions flow essentially from the definition of natural series, that is, they *follow* the number concept, do not precede it.

The axioms IV and V evidently originate in the same definition.

combinations of these groups similarly formed of their component objects.

It is *à propos* to take up here the opinion of Helmholtz that our number concepts have at their basis order-symbols. I have shown that the primal material for the number concept are the sensations of co-existence. These do not in the least contain impressions of order. Besides, I dare say, the idea of numerical order must be preceded, even if in only a half-aware way, by the concept of natural series.

The above example of 3 + 2 pebbles vis-a-vis 3 + 2 apples is not an argument *ad hominem*, for one could not define the idea of sum clearer and in a more natural way than Helmholtz does.

Evidently there are hidden premisses for this proposition, too. The axiom of independence is one.

But the permutation and combination laws clearly originate in the sensual material of number concepts. The very impressions of divisible co-existence are marked by the sole trait that objects in a given group are perceived at the same moment but as distinctly separate. When the latter two conditions are present, neither the character of relative position of the objects, nor their nature can alter the resulting impression of divisible co-existence.

It is clear then, that the idea of sum, and therefore, all the concepts, axioms, propositions and other elements of arithmetic, all those deducted from the concept of addition, can have neither purely hypothetical meaning which the empiricists like Mill wish to imply, nor the sense of order accepted by the mathematicians.

These elements of arithmetic are ingrained in the primal material of number concepts.

VII

Thus far my *exposé* tried to show that all elements of the system of arithmetic bear the formal character of its premieval psychological material: the pure sensations of divisible co-existence.

The latter are roots of primal counting forms: the number images. Owing to their absorption by our awareness minding its own sensation of halts, the concept of number emerges with the attending idea of unit and other number contents.

The key term of equality is applicable to groups that conform to conditions of countability, that is, to number images only. A premiss of the basic propositions of arithmetic, the axiom of the number's independence from the nature of objects which are being counted, is nothing else but the order of co-existence, the law of pure relationships of divisible co-presence.

But such relationships express the nature of outside experience as a whole, not of its parts (see ch. II). Their law must therefore speak of conditions for the outer experience itself: of the terms of its structure, not its perception. The axiom of independence covering also the law of co-existence and ruling the contents of arithmetic, is an issue of the transcental subject, bearer of forms and conditions of outer experience as a whole: not of the empiric subject; not of the one recording sense data.

Later on, I shall try to show that the nature of arithmetical judgment can be explained on the strength of the above proposition. For this one has to fight with either: flagrant empiricists who refute the *a priori* nature of arithmetic, and Kantian psychologists who accept the latter.

With the empiricists I agree on the number concept originating in outer experience, and disagree as to what aspect of sense data those perceptions present. With the Kantian psychologists I part ways on the question of the physical existence of numbers, on the establishment of facts, that is, but agree – as it will be seen later – with their conclusion as to imperative certainty of arithmetical truths.

In these animadversions the oddest position is that of Kantian psychologists. Wishing to sustain the Kantian doctrine in epistemology, but unable to square it with their viewpoint as psychologists, they are caught in a net of *professional* errors which could be called "psychologisms."

The root of these errors lies in the psychologists' seeking the source of the universal applicability of mathematical propositions in the action of the empiric subject, in the course taken by psychical functions.

When Siegwart and others try to draw the nature of arithmetic from the reflexion on the halts of consciousness (action of an anthropological factor), and define axioms as logical premisses of experience; when Kant deduces these traits from the analysis of judgment thinking that the nature of arithmetical truths cannot be figured out unless through the idea of analytical and synthetic judgments, – they repeat the same "psychologism," a distinct philosophical error.

The source of such traits in arithmetic must dwell outside the action of the subject receiving sense-data: outside acts of speculation and of reflection on its own halts.

Q

A savage, a child in the domain of small numbers, a man of culture in the realm of large ones, "distill" the number concept from what is but vague, half-sensual images, when they see that number is composed of units in the same way as a group of things is composed of one.

To this mental process particular details are added. Every single arithmetical judgment is accompanied firstly, by a certainty of its truth and validity for every personal experience; secondly, by a still subtler realizing that number is a formal trait, one not arising from the inner qualities of things being counted. To such awareness only can a certainty be linked that number is composed of units as a group of things of one.

I shall devote the rest of this essay to the source of the above mentioned particularities.

α) *Universal applicability.* Three viewpoints rule this problem: those of psychologists-empiricists (Mill), of the mathematicians (Helmholtz and Stolz), and of the Kantian-psychologists (Siegwart).

Mill insists that "number enjoys universal applicability because its nature is concrete" and that "if algebraic truths are right in relation to all things, it is no wonder that symbols evoke in our mind no idea of some definite and private things." [1]

But the question is precisely this. How is it possible that something of concrete character should hold such applicability?

If number is concrete in Mill's sense then we could know that three apples *plus* two apples are equal to five apples; or three pebbles *plus* two pebbles make five, and yet have no assurance

---

[1] John Stewart Mill: *System of Logic*, Vol. I, Chapter VI.

that $3 + 2 = 5$.[1] If number is but a concept abstracted from physical qualities of things, it would not be thought of otherwise than in a physical sense and would evoke ideas of private things.

Abstract arithmetic would have no sense if its concepts were linked to outer experience alone: just as addition of denoted numbers of different denominations has no sense.

Of course, a physical arithmetic is possible, one resting on a limited application of the idea of equality and so on, and permitting crude operations of physical joining and detaching of groups, a primitive kind of addition and subtraction (see ch. III). But such arithmetic would be lacking the axiom of independence and the affiliated traits.

This is what drastically separates abstract arithmetic and physical. He who says that the latter was arithmetic of the lower grades of culture and formed the roots of abstract arithmetic, should be reminded of the delimitation of the anthropological and philosophical problems (ch. I).

The mathematicians' viewpoint is much better sustained.

Applicability, as well as certainty of abstract arithmetic rests on a transfer to it of the concepts of equality and operations formerly set up for the empiric material: those of number images such as *Vielheit* (multitude, many) or *Anzahl der Dinge* (number of things).

But giving this setting, say in Stolz's presentation, a closer glance, we perceive again one purely formal premiss. If in two groups (multitudes) each element of one group has a corresponding one in the other group, a permutation of some members would not change the situation. This proposition is, in essence, the axiom of independence itself. To base the transfer on this axiom means only a renaming of physical arithmetic into abstract, and bridging the wide gap between the two in one jump.

Finally, the Kantian psychologists say that universal applicability of number concepts flows from their origin in the reflection of consciousness over its pauses. The following speaks very clearly of the viewpoint of this school:

"In number relations we are confronted not with changing phenomena of the material world but the inalterable laws of our

[1] Criticism of the usual opinions in this matter of the evolutionist theory will be found later on.

mind ... When we utter an arithmetical proposition, we take it that every other thinking individual has a consciousness capable, like ours, to interrupt itself in order to perceive number relations ... We base this thought on conviction that those halts of awareness are effected in the same way by all individuals organized as we are." [1]

Here we find the same error, the "psychologism" which tries to explain by an acting psychical function (the empiric consciousness) something cradled in transcendental conditions.

Why do the laws of our mentality or awareness have greater stability than the laws of the material world? Does the applicability of arithmetic derive from sheer certainty that the consciousness of all men is similarly organized?

The latter consideration is feeble. An empiricist could just as well say that the universal applicability of number, in itself an abstraction from physical qualities, is based on all men's mode of sense-perception, on one similarly organized. If number be but an abstraction from processes of awareness, its applicability would differ from that of abstraction from the physical traits of outer experience in volume only, not character.

The origin of number in reflection explains its applicability also to the spiritual aspects of experience, such as the psychologists' interpretation of the feeling of certainty. Then all concretely possible application of number to the entire contents of our experience becomes manifest; all but the abstract, the *feasible* application – to every possible experience.[2]

Number has universal applicability neither because it is concrete, nor because it is abstract. Like space, number is neither sheer empirical fact, nor pure abstraction. It is a transcendental function, for it speaks of the nature of experience as a whole. The very right of unlimited use of arithmetical truth must lie in the immutability of the order of arithmetical contents. This order amounts to the law of co-existence, the law of experience as a whole, the law of the very organization of experience, and therefore, is inherent in the group of transcendental conditions of

[1] Prof. Tchelpănov: "The a priori elements of consciousness" (*Problems of Philosophy*, Moscow).

[2] Strongest argument against this thought would correspond to that of new geometricians refuting the universal applicability of euclidean geometry. The answer to such argument can be seen here (in ch. IV).

experience nestling iú the transcendental subject, builder of outer experience.

β) *Absolute certainty,* Mill affirms that the absolute truth of mathematics flows from the conformity of its judgments with the hypothetical premissess. Were we to take it that the axiom $1 = 1$ is a proposition not entirely certain, arithmetic would become a science as little sure as physics.

But this is not so! Arithmetic would not become then a science of little certainty: it would lose meaning altogether. Which shows that the truth of arithmetic is tied to its abstract essence; that number has a stronger title to universality than sheer physical feasibility; that the necessity of the laws of physical arithmetic is not the highest measure of necessity demanded by outer experience.

Besides, are the impressions of certainty typical for mathematical propositions of the same nature as those of conclusions correctly drawn from hypothetical premisses? Does "the necessity of mathematical truths" really lie only in that they inavoidably follow the suppositions from which these truths are deduced? . . .

No, such a necessity is the absolute belief in the conformity of arithmetical truth with every possible fact: not the logical necessity of its conformity with premisses.

Mathematicians base the absolute certainty of arithmetic on complete induction, on conclusion from n to $n + 1$. But this is a mere renaming of a philosophical question.

A much subtler enemy of our convictions is the evolutionary school. Really, if ontogenetic development or personal history cannot be the source of arithmetic's nature, then philogenetic unfolding *is*, perhaps, this source. For the latter's almost infinite duration, the experience of the species has turned the possibility of truth in arithmetic into certainty.

All of this would be right if, in ontogenetic life, the transition from perception of number relations to the awareness of their truth were not a leap. But such occurs in every personal unfoldment. This means that it is enough to direct our consciousness in a certain way to make the entire cycle of experience available to the individual.

This is possible only because the empiric subject "coalescing" in some way unknown to us with the transcendental, conditions the very order of experience and becomes aware of this highest of perceptions as soon as our awareness is intent upon the truth.

The *a priori* nature of arithmetic lies in that no new experience can broaden its concepts; for the latter are void of the material contents that are called experience. The experience necessary to arithmetic is one *a priori* created. This is arithmetic's main difference from the nature science.

The nature science's truths have no absolute necessity not because one can not achieve ideal measurements. Had we instruments of faultless precision, the certainty of truth of a scientific proposition would be equal to a fraction, where the numerator is the number of cases confirming the truth in question and the nominator – the number of all equally possible, non compatible (in relation to each other) and the only possible cases of experience whose contents is the fact being described. This certainty would be $\frac{a}{\infty} = 0$, which means: there is no certainty.

In arithmetic where experience is created *a priori*, we know beforehand all equally possible, noncompatible and the only possible cases of that special experience which sustains the given proposition. Therefore the certainty of an arithmetical statement is $\frac{a}{a} = 1$ (a is the number of cases perceived).

Which means that we have here full certainty.

VIII

Truths expressing the nature of the outer experience *in toto* are *a priori*, for they are linked to an awareness of the very frame and fabric of experience. Arithmetical truths contain that trait of awareness. Their object consists of the most general and formal qualities of any experience; their root, the axiom of independence, presents the conditions of the outer experience's structure: the law of co-existence.

Arithmetic's *a priori* essence is thus a consequence of the transcendental *a priori* nature of arithmetic's axioms; of something predating sense data; not logical or psychological *a priori* – suppositions preceding experience.

The axioms of arithmetic precede experience potentially not as its logical premisses (as the viewpoint of Kantian psychologists has it), or as expression of the nature of outer experience (as the psychologist of the empiric school think). These axioms are possible and exist *together* with the experience as conditions for the latter.

The particular point of my own views I can express in this juxtaposition.

An eminent thinker of the past said:

"Number is an *a priori* concept: for it does not depend on sense data, is not conditioned by them; number relations are not proved by outer experience and are not refuted by it."

I say:

"Number is an *a priori* concept even if dependent on sense data, even if conditioned by them; even if number relations are proved by outer experience or are refuted by it."

Critique of New Geometrical Abstractions

To be discussed here is only the general meaning given to new geometry – to its abstractions or generalizations.

New geometry presents a system of ideas interdependent and logically compatible. Thus my task allows to view the details of new geometrical doctrines as evidence looking toward one main thought.

But in its application to the general question – of the *a priori* nature of geometry's axioms, new mathematics became material for epistemology or, in the older term, gnosticology. And the whole problem of value, meaning and power of metageometry's reasoning or conclusions is to be solved by correlating them with the ideas and results of epistemology and, therefore, also with the basic facts of psychology.[1] In his essay on the *Foundations of Geometry*, Bertrand Russell is doing just this. But he does not see or, at least, speak clearly, of the relationship of mathematical feasibility to the epistemological. Russell's reasoning does not convince us that the former is the deciding factor.

I

As for metageometrical generalizations and abstractions, their general significance lies in being ground for a denial of the *a priori* nature of axioms that pave our orthodox geometry. The argument shapes itself this way.

If systems of geometrical concepts and space continua different from the euclidean can be construed by our mind – different yet "on the par" with orthodox geometry, that is, free from inner contradiction and similar in structure, – then our

[1] I have shunned any explanation of elementary detail in dealing with new geometrical concepts. This would have interfered with my task.

euclidean geometry and space concepts are not imperative.[1] For brevity's sake, let us call this proposition our *main argument* or that of feasibility.

It takes on various shades caused by change of meaning in either the premises or conclusion.

To begin with, the argument of feasibility takes on another guise when the query as to the "truth" of geometry is shifted to the "convenience" of its application to space (Mach, Poincaré *et alii*).[2]

The latter school tries to solve the problem by abolishing it. Yet the proposed change does not discard the question as to where lies the source of our assurance of the truth of our geometrical ideas.

This belief is covered by neither the contents nor the nature of our physical experience. And if the above main argument explains the truth of our common-use geometry by the maximum of its applicability, we are offered here, in fact, the reset theory of John Stuart Mill on the hypothetic applicability of mathematics to experience – a theory many time challenged.[3] An essential criticism of this theory of applicability can be found here.

One of the bases for the above argument lies in the projective concept of distance. The latter depends on the choice of two points which, together with two given points, yield the needed anharmonic relation. Distance appears then to be a variable complex, and the choice of its frame depends on the convenience of application.

These propositions and the particular idea of distance underlying it, were neatly demolished by Bertrand Russell. He has shown that the projective idea of distance – the positional relation of four points – not only does not eliminate the metric idea of distance but presupposes it.[4]

The sequel to the argument of feasibility is being altered also this way. *Not all* of "our" ideas of space are imperative if other

---

[1] Compare this with Riemann's discussion in *Ueber die Hypotesen, die der Geometrie zu Grunde liegen.* See Riemann's *Gesammelte Werke*, pp. 272–273.

[2] In his essay *Sur les hypotèses fondamentales de la géometrie* Poincaré compares the choice of geometry to be applied to "*our*" space with the selection of co-ordinates. This view influenced by Avenarius, Poincaré calls a commonplace which he would not discuss.

[3] Lazare Saminsky: *The Roots of Arithmetic.*

[4] Bertrand Russell: *Essai sur les fondements de la géometrie*, 37.

geometries are possible. Only axioms common to euclidean geometry and non-euclidean are marked by necessity. It is imperative, for example, that space should have a definite number of dimensions but it is not necessary that there should be three dimensions.[1]

This correction does not alter, however, the essence of the main argument or the nature of ideas linked in it. The meaning and power of this argument would not change if its conclusions were related to even one trait of euclidean space or to one euclidean axiom.

But aside from this, the above correction contradicts the ground facts of epistemology.

There is a seal of assurance on the "minor" axioms of geometry, a no lesser conviction than that marking general axioms. Euclide's postulate on parallel lines, of his axioms the one most limited in application, has in our minds a degree of assurance no lesser than that of general axioms, such for instance, as the necessity for a definite number of dimensions.

This is why a correction in our main argument such as I have just quoted, can not be. If even one axiom is not imperative because of our "argument of feasibility," then none has necessity.

Thus, the defense of the empirical nature of the axioms – in as much as the basis for such belief lies in merely mathematical abstractions, must be led by the clear cut, direct argument formulated at the opening of this essay.

Our argument of feasibility has, however, three variations dependent on the interpretation of the equality of title to existence for the various frames of space.

Geometrical systems of sheer logical equality, that is, sourced in the proposition that they are equally possible, are the basis for the argument of pure feasibility. Geometric systems of an analytical equality, those having similar and logically compatible contents, feed the argument of feasibility illustrated analytically. And finally, systems with a gnosticological (or epistemological) equality, those allowing an imaginable geometrical interpretation, support the argument of concrete possibility.

---

[1] Russell, *ibidem*, pp. 143–179, shares this view.

α) If the abstract feasibility of other space frames, if the mere thought of their chance for existence could serve as an argument against the *a priori* nature of euclidean space, then, in general, no absolutely sure *a priori* judgement would be possible. No matter how powerful the argument, sheer abstract feasibility of an opposite thought would be proof that our present judgement is prompted by experience.

One can, however, fancy systems built on mere negation of the necessity of euclidean concepts: built, that is, on a pure feasibility of space formulae that negate this or that trait of *our* space.

Feasible seem indeed systems of space with changing curvature. Bringing the abstract negation of our common experience to its logical end, one could think of space systems where lines are not formed by a point in motion. One obtains such a system by basing it on the existence of a function that allows for no differential, the Weierstrass function, for example, which sums up a certain progression $\sum_{n=0}^{\infty} b^n \cos (a^n \times \Pi)$.

Here x is the concrete, irregular variable, a — a certain whole number, b — a positive constant less than *one*-, and all of this conditionned on $ab > 1 + \frac{3}{2} \Pi$.[1]

It is possible to formulate spaces as groups of ∞^2 positions of curves defined by this kind of functions. In such spaces, because of the Lagrange theorem, not every position would allow for a tangent.

It will be shown later on that from an epistemological point of view such frames of space are absurd. Their admissibility and logical equality with the euclidean frame of space reduce themselves to a mere stating of the general source of logical feasibility: "any proposition is either accepted or rejected."[2]

But can something built on an abstract negation of a certain proposition deny the latter's necessity on the strength of a sheer logical construction?

Abstract possibility of other frames of space *can not by itself* serve as an argument against the necessity of the euclidean system. The argument of pure feasibility has no more concrete proving

[1] Killing: *Einführung in die Grundlagen der Geometrie*, Band I, s. 172.
[2] In traditional logic – the law of the third excluded.

power than the very existence of the logical disjunction: for abstract possibility *is* this disjunction.

β) The main argument can be illustrated by an analytical picture of the possible systems of geometric relations similar in structure and logically compatible. Such illustrations are taken for a proof of equality of possibility – for such geometries.

Thus distance, the definition of which lies at the bottom of geometry, is defined by functions of differentials of co-ordinates in any space of constant curvature: such is, for example, the equation for the segment of a curve in four-dimensional space $ds = \sqrt{dx^2 + dy^2 + dz^2 + du^2}$ similar to the well known formula in *our* geometry of *common use.*

Equations figured out for the flat pseudo-spheric triangle differ from the equations for spheric triangles in this only: the mathematical figure for their sides includes an imaginary multiplier. They change form only when the function of the angle of parallel relationship is entered (for instance, $\sin A : \sin B = \cot \Pi (a) \, i \cot \Pi (h)$; and hyperbolic functions, as well. [1]

In general, eliminating from the ensemble of euclidean axioms that of infinity of a straight line and the postulate, one finds it possible to construe trigonometric systems of space with a constant curvature. Such frames of space could belong to any of the three types – the elliptic, parabolic and hyperbolic. [2]

But how far goes the proving power of the mere feasibility of such analytical pictures of possible geometries? A proposition which, broadening a mathematical concept, removes from under it the soil of experience coming from the senses, admits more than one illustration. Yet presumably, only one such figure should be possible: only one correlated with the abstract formula.

The analytical picture of relationships of complex quantities is illustrated by the geometric theory of vectors. But will anyone argue that this correlation is imperative, is not "managed" – as any "building" of algebraic formulae is?

Once illustrated ideas have no corresponding facts in ex-

---

[1] Beltrami: "Saggio di interpretazione della geometria non euclidea," fourth section (*Giornale di matematiche*, 1868, Vol. VI.)

[2] This manner of geometrical lay-out can be seen in Killings book *Einführung in die Grundlagen der Geometrie*, B. I, § 24.

perience, the choice of illustrations is bound to be more or less arbitrary.

At the first glance, this argument seems to be fenced off by a remark made by Lobatchĕvski for another case.

As it is well known, pseudospheric geometry illustrates its analytical results by a revealing of some quadratures, that is, summations of definite integrals. Lobatchĕvski remarks that application of geometry of common use to the analysis of infinitesimals has the same meaning. [1]

True, of course! But aside from the fact that the integrals squared by Lobatchĕvski's geometry are "selected," the application of common use geometry to analysis is also built on conventions, those of analytical geometry. This means, that here too, the selection of an illustrating example contains an element of choice.

Therefore, the feasibility of the above analytical pictures shows a mere chance not of other geometries, but only-other systems of formulae which are non euclidean. This fact does not *by itself* sustain the main argument.

The logical compatibility of such analytical ensembles has no greater proving power, either.

This compatibility is formulated in the classical argument which, linked to new geometrical concepts, shows the independence of Euclide's postulate from his other axioms. If a given system of concepts X can be shown to lead to conclusion Y and, at the same time, to a directly opposite conclusion Z, then the systems XY and XZ contain no contradiction, and neither proposition, X nor Y, is indispensable.[2]

Lobatchĕvski, for example, defines analytically the function of the angle of parallel relation by the equation $\cot \frac{1}{2} \Pi (x) = e^x$; then deduces his equations for pseudo-spheric triangles. And finding these equations in accord with the conditions for the triangle's existence (with the system X) prooves in this way the legitimacy of his geometry.[3]

But the harmony of the system xy where y is the equation $cot \frac{1}{2} \Pi (x) = e^x \neq 1$, one belonging to hyperbolic geometry, with

[1] Nicolaï Lobatchĕvski: *Complete Works*, Volume I.
[2] See Lobatchĕvski, *l.c.*, p. 67; Killing, *l.c.*, B.I.
[3] "Lobatchĕvski: *Imaginary Geometry* (*Complete Works*, p. 76 etc.)

the system xz, where z is the equation $\cot \frac{1}{2} \Pi (x) = 1$, that of parabolic geometry, has its source in a premiss that substitutes the idea of geodesic triangle for that of straight-line triangle. This substitution also presupposes that through a point outside a given straight line it is possible to pass more than one parallel line, and also – none.

The unfolding of such premisses is but a variant of the disjunction:

$$\cot \tfrac{1}{2} \Pi (x) = 1 \qquad\qquad \cot \tfrac{1}{2} \Pi (x) \neq 1.$$

Here one finds embodied, that is, recounted in systems of formulae, everything logically feasible in space relations. The compatibility of analytical structures like this, adds no power to the argument of pure feasibility; makes it no less of a sheer abstraction.

γ) For a *concrete* possibility of happenings, being merely imaginable is not enough. For the feasibility of a space this is sufficient. Therefore, if mathematical generalizations could show the possibility of space relations unlike euclidean but equally legitimate in gnosticological respect, the main argument would acquire real strength. Close to such goal is the famed discussion of Helmholtz in connection with the results reached by Beltrami; and also, the description by Helmholtz of the sensations of an observer with *our* visual perception who happened to find himself in a pseudo-spheric space.[1]

Certain repartees made to this strongest reasoning of new geometry, one to proove that non-eudiclean spaces are imaginable or can even be visualized, have pretty much uncovered the weakness of the latter argument.

Helmholtz's picture shows that objects in space, curved or flat, stand in the same relation that is established for figures on a curved and flat surface by the well known problem of Beltrami: to transfer points of a curved surface to a flat one in such a way that the geodesic lines of the curved surface correspond to the straight lines on the flat one.[2] The possibility of visualizing relations thus construed is considered by psychologists a proof of

[1] Beltrami, memoir "Saggio di interpretazione", *ibidem*, p. 287, etc. Helmholtz: "Ueber den Ursprung und die Bedeutung der geometrischen Axiomen" (*Vorträge und Reden*, Band II, s. 26, etc.).

[2] Here one has to translate non euclidean formations into those of common use.

"imaginability" of space relations unlike the euclidean, even if it is impossible actually to visualize – or in some way imagine – non-euclidean spaces.

The meaning and the prooving worth of the analogies that show in the above correlations, analogies built on an abstraction merely mathematical, will be discussed later on, in chapter III. But, apart from this, Helmholtz's description does not really sketch totally unknown frames of space.

Let us imagine beings born on and living on a globe or in a spheric space. In spite of the propositions of spheric geometry, such beings would have an idea of parallel lines. This role would be played by every geodesic line with its correlated, equidistant curve: in our common vision, by the arcs of geodesic and spheric circles obtained from dissecting of the globe with parallel flat surfaces. In pseudospheric space this role would belong to *its* geodesic lines with *their* equidistant curves.

True, one must note, that Russell refutes the possibility of presenting non-euclidean space relations in terms of ones germane to the euclidean continuum.[1]

My argument, however, does not lose power even if accepted conditionally. Either non-euclidean space relations in some way "translated" into euclidean figures make sense: then my argument is valid. Or Russell is right; which means that non-euclidean space can not be imagined, or in any way visualized.

Beings born and living in spheric space would be aware of ideal lines (those not changing direction) which always cross. But they also would be aware of other lines that do not cross and are in the nature of our parallels: that is, given a straight line and a point outside it, we can find only one other and equidistant straight line passing through the point. The same is valid for the pseudo-spheric space.

The above described inhabitants of spheric space would say that all the trajectories which present the location of shortest distance segments cross in two points-about the same visual fact which we describe as the arcs of the grand circles crossing in two points. They would insist that a straight line and its equidistant do not cross about the same visual fact which we describe

---

[1] Bertrand Russell: *Essai sur les fondements de la géometrie*, p. 68.

thus: a geodesic line and spheric circle obtained from dissection by parallel flat surfaces, never cross.

It is clear then that beings living in a curved space would have only a *geometric* idea of parallels other than ours, for this idea pertains to the relation of two shortest distances between a twin pair of points. But the sensual content of parallels is a perception of relationship of lines equidistant; and this would *not* be different from *ours*. Therefore, Helmholtz's picture shows the abstract possibility of beings with a non-euclidean train of perception, or rather, with non-euclidean relation between the objects of perception: that is, with just another geometry, not with a completely different psychic instrumentality of perception.

Our thought on the matter leads even to a result totally unexpected by mathematicians. The latter wish to prove that the same basic geometrical abstracts (the straight line, parallel lines, etc.) could be related to differing geometrical media; that different sense data could be linked in one generalization, i.e., one formal geometrical concept. But it turns out that, just to the contrary, the same geometrical images can be correlated with strongly differing geometric concepts.

Thus the figuring out of the Beltrami-Helmholtz pseudo-spheric relations within a full sphere of euclidean space does not sustain a possible "visualizing" of other space relations than "ours." The proof of such "imaginableness" is reduced to merely this: given *our* space perceptions, one could fancy geometric designs that would be normal for a different train of space relations.

Let us take as an example the geometric locus of straight lines parallel to a given flat surface at a given point of pseudo-spheric space. In our imagination such geometric place would look like a conus with curved radii which a parallel flat surface approaches assymptotically.

Such a formation alters only images and concepts considered "normal," not our space relations. Therefore one can not agree with even the "softer" opinion that, "with the aid of Helmholtz's method we can reach an understanding of pseudospheric space relations, but obliquely, through a translation."

But let us, for the sake of argument, suppose that the ability to imagine other space relations is proven. Does this mean a concrete possibility of non-euclidean space?

Such possibility could be indicated by feasible changes in the *qualitative* traits of our perception; by a feasibility of changes not in the measurable, but in the indissoluble nature of space form.

What is to be proven, is not the ability to "imagine" objects with inner measurable relations other than euclidean, but the concrete possibility of a new order of perception. We construe objects in non-euclidean space by juggling or curving linear measurements of euclidean objects. We can not really figure out non-euclidean space in the sense of a new order of perception.

To affirm the possibility of non-euclidean space by conceding the prooving power to mere feasibility of altered euclidean relations – is to repeat the error of a mathematician who mistakes a variation of some function (integral) for its differential.

If a given system of conclusions is right within the limits of the differential of a function changing its quantitive import with the change in its variables (in my analogy, with the change in space relations), these conclusions may still be inapplicable to a variation of a function where its very nature is altered (in my analogy – where the very form of perception is changed).

Thus, the argument of *concrete* possibility of non-euclidean space is not sustained by new geometrical concepts. Their analytical contents embody the logically predictable, but only in systems of formulae.

Which means: the arguing power of metageometrical abstractions dwells in nothing more than a sheer presuming of non-euclidean space.

The true meaning of this argument and its restrict validity have been shown. It remains to bare the epistemological root of this debility.

II

If the main argument repeats in its premisses but a certain logical disjunction, why should this fact deprive the argument of its power; why does it limit this argument's applicability? This query is the driving point for the following.

The mathematical concept of space is embodied in a knot of geometric axioms and definitions.

The metageometric generalization of space widens the concept by eliminating characteristics proper to only one type of geometric relations. Handling concepts not derived from experience, the above elimination of space characteristics has no limits compelled by mathematics. When the denying of necessity to euclidean space-symbols reaches the widest mathematical abstraction, the idea of space loses its contents (space with absolute positions). We arrive at an infinite number of continua logically and analytically feasible.

But ... if such a logical end of the generalization of space brings one to a psychological *reductio ad absurdum*, that is, to the impossibility for this order of space – a would-be space with variable curvature – to serve as a form of perception ... if on the other hand, out of all possible *continua*, mathematical generalization chooses but those of an "orderly" train, in fact – only those of constant curvature, there must exist a limit to geometrical abstraction. Or, at least, this abstraction can neither be absolute, nor extend *ad infinitum*. Apart from logical consistence, geometric abstraction must bow to yet another demand.

I will show that this demand or condition is *gnosticological sense*. Space abstraction must make sense in regard to philosophy of knowledge itself.

There must exist a limit inherent in the generalizing, in the abstract refining of the mathematical definition of space. There should exist an absolute core of geometry: the minumum of determinants common to all possible space systems. And these must present traits without which the very idea of space loses meaning.

But is there such absolute minimum, in a purely mathematical sense?

The ensemble of even the most general axioms of projective geometry does not constitute such absolute core. In Russell's summing up there are three such axioms.[1]

Axiom I. "One can distinguish different parts of space, but they are all alike qualitatively, and they differ only by the immediate fact that they lie one outside the other."[2]

---

[1] Russell: *Essai*, Chapter III, Section A.

[2] I am translating this from Russell's French version of his *Essay on the Foundations of Geometry*.

But the axiom of "space-splinters" lying one outside the other, does not clearly define space as something quite different from other *continua*, time, for example, a one-dimensional expanse. The pure form of time, too, has no other distinction for its moments save that they are lying outside each other.

Infinite divisibility of space (axiom II), comprising points, zeros of distance, belongs also to time. As a result of the latter's division we obtain moments, zeros of time.

Axiom III – two points define one figure, the straight line; three points – a flat surface, etc., – belongs also to metric geometry.

In the latter, its absolute core consists of three axioms: free mobility; the axiom of distance or straight line; and the necessity for a definite number of dimensions.[1]

The first of these axioms is not general in a mathematical sense. Spaces where this axiom has no validity, are feasible and even imaginable: as geometric loci of surfaces with changing curvature.

The axiom of straight lines is not valid for geodesic lines of the sphere, when they are drawn between polar points.

Felix Klein shows that this fact is not necessarily connected with the sphere's positive curvature.[2]

As to the axiom of dimensions, given the non-generality of the above first two axioms, the former's substance defines space only as a function of a definite number of variables. This in itself does not divorce the concept of space from other definite functions, to say nothing of the fact that the idea of dimensions itself hangs on conditional points (see chr. III).

Therefore the axioms of projective and metric geometry are either not indispensable, mathematically or generally; or they possess a generality of a kind which precludes a defining of traits purely geometrical.

---

[1] Russell, *Essay*, Chapter III, Section B.

Helmholtz presents these axioms as *admissions* indispensable to the forming of the idea of space, but he alters the formula for the axiom of distance. "One admits the existence of unalterable systems of points, that is, between 2n coordinates of every pair of points on a hard surface there is an equation which does not depend on the body's movement and is the same for all coinciding pairs of points."

Helmholtz: "Ueber die Thatsachen der Geometrie" (*Wissenschaftliche Abhandlungen*, II Band, s. 622).

In addition to the three axioms mentioned, Helmholtz defines also the axiom of *monodromy*; but this is essentially covered by the axiom of free mobility (See Sophus Lie's remarks on the memoir of Helmholtz).

[2] ,,Ueber die so genante Nicht-Euclidische Geometrie," *Mathematische Annalen*, B VI, s. 125.

This is why, in a mathematical sense, an absolute section of geometry does not exist.

Pushed to its ultimate logical end, abstraction reduces formal geometry to nothing. One will arrive at spaces where no formula of equality or homogeneity will work; where equation based on symmetry or sameness of measurement is not possible.

The following will show that the absolute core of geometry that sets limits to geometric abstraction based on pure feasibility of non-euclidean relations, is defined by traits gnostological and psychological, not mathematical: that is, by the ability to formulate the law of space perceptions.

I, for one, deem wrong the following opinion of a noted Russian philosopher, the late Professor Tchelpănov: "It would be fallacious to think that the proof of *actual imagining* of other space relations in the same way as we visualize *our* space, is the task of metageometry. It would be enough for the latter to prove the sheer "admissibility" of other space relations." My thesis is that for the *epistemological* aspect of the problem which new geometry has undertaken to solve, mere "admissibility" is not sufficient.

This is why the absolute core of geometry contains all the axioms of metric geometry quoted above, in spite of the fact that not one of them is *mathematically* imperative.

The psychological meaning of the sense of space or expanse is given, in its essence, in the well known theories of the nativists and genetists. The former – Stumpf, James, Hering – tried to prove that the sense of space is primal and integral; cannot be reduced to more elementary sparks of sensation or related to the latter's mechanism. The genetists – Wundt, Lipps, Helmholtz – tried to refute this doctrine.

This difference of viewpoint was not always strictly adhered to.

One of the facts, for example, that speaks in favor of the "irreducibility" of space-sense, namely, the existence of a purely optical difference in very small objects when viewed by the eye completely motionless, is recognized even by the English school, partisan of non-primacy of the space-sense.

In the same way Lotze who is of the genetic point of view – in the problem of objectivation admits the primal nature of space sense. He agreed with the nativist Stumpf who proved that "in the very sensations their order must be given as an immanent quality." But pointing to the fact that we can say precisely which ear has received a certain sound sensation, Stumpf used to insist that the sense of space is the immanent moment not only in visual sensations.[1]

The genetists counter the nativists' discourse with strong critique – their only real vigor, in my opinion. Their own positive theories are vulnerable.

Thus, Wundt's doctrine sketches the process of visual perception as a confluence of local signals of the retina and two-dimensional continuity, with the system of signs signaling sensations of motion. Owing to a gradualness in muscular feeling the latter present a one-dimensional continuity, as Wundt maintained.

Perception of space is thus a measuring of the ensemble of the retina's signals with symbols of motion.[2] This theory was demolished by the experiments of Dove and Hering who showed that in their substance visual sensations do not depend on the eyes' muscular movements. Hering's experiments present a strong argument for the nativists' point.

The genetists, on the other hand, efficiently destroy the notion of identity of the retinas by which the nativists aimed to explain the single-object vision. Against this theory the genetists pit the phenomena of binocular conflict of colors and sequels of the operation on the cross-eyed. When, however, the genetists – Helmholtz, for example – try to use the same facts to prove that perceptions of space – in this case, the fact of single-object vision – are explained by muscular – tactile experience, and that each of the two eyes experiences a different fixation of objects, Stumpf, for one, shows that from the binocular conflict of colors it is possible to draw an opposite conclusion, and be as persuasive about it.

There is still another bright fact that proves, in my opinion, that it is impossible to reduce perceptions of space to some

---

[1] Stumpf: *Tonpsychologie.*

[2] See Wundt: *Logic*, Volume I, and also his studies in psychology.

elementary psycho-physiological processes. In spite of the non-continuity of the light-perceiving elements of the retina, in spite of the existence of the blind spot, the field of vision is continuous.

Explanations advanced by genetists–Wundt and Helmholtz – are not satisfactory. Even if the thesis that the continuity of the field of vision can not be explained psychophysiologically is open to question, one has to bow to the opinion of Lipps that "filling the gap in the field of vision corresponding to the blind spot, presents but a case of the pervading space sense." Thereby the question of filling the blind spot is the same as that of continuity of space compatible with the interruption of sensation.[1]

In the solving of the latter problem lies, in my opinion, the end of the dispute on the derivative nature of the space continuum.

The antinomy existing in the general problem, through which the genetists made it very clear that spacial quality is not as immediate as experience of our senses, and the nativists showed with no lesser eloquence that space can not to be reduced to other elements in our psyche, – this antinomy must, indeed, prevail also in the problem of the dissoluble nature of the dimension of depth, or of the third dimension being a complex, not primal quality.

Genetists, such as Lipps and Berkeley, tried to prove that the perception of depth can not be a direct, immediate sensation. Their reason is that, to appraise depth, the eye must see both the object (or the point) and itself.[2] Lipps takes an example of two objects, of which the farthest is only partially screened by the nearest, so that the visible part of the far object seems to be on the same flat surface with the near one. Therefore, says Lipps, there is no perception of depth: only knowledge or conviction resulting from the opportunity of appraising the distance-measure of depth at other circumstances.

Lipps' psychological argument for the impossibility of direct perception of depth is that the sense of depth would be that of the void between objects, of something which has no spatial nature.

But again, the nativists' critique is no less plausible. *Their*

---

[1] Lipps: *Psychologische Studien.*

[2] This was refuted by Cornelius saying that the eye compares the relative position of the object and of another point of the body it belongs to, which the eye observes. But in this case the eye perceives not depth but an ordinary straight line on a flat surface which it observes sidewise.

argument against the derivative nature of depth is that the genetists mistake flat surfacial expanse for reality. Dimensions themselves are nothing but abstraction.

Psychological critique by the nativists shows that it is impossible to deduce the third dimension or the objectivation of our perceptions from psychophysiological elements: such as modifications of flat surfacial expanse, auxiliary muscular moments, etc. Even those psychologists who are to some extent favorable to the genetic point of view, are inclined to think, when in critical mood and chary of bold decisions, that muscular and tactile sensations as well as other auxiliary moments are only partial, not the main conditions for the sensation of depth.

Of course, the nativists too, have their positive arguments. In spite of their opponents' telling analysis, they manage to explain very well, indeed, through even a mere describing of the sensation of depth, that this quality is a directly perceptible, indissoluble trait of the nature of space.[1]

The analysis of space perceptions convinces us that the series of facts and principles unveiled by the two schools of thought, are compatible.

The primal nature of spatial quality with its directly felt sense of depth, the immediacy and prime nature of the process of objectivation of sense-data,[2] in fine, the impossibility of reducing the nature of space perception to simpler psychophysiological elements, is perfectly compatible with the proposition that the very existence of depth is shot with conflict and can not be explained psychologically; that space perception is not a sense datum of the same order as any other sensation.

How is such a concord possible?

The facts, that special order itself and the process of objectivation of our sense-reaction not reduceable to supplementary

---

[1] William James: *Principles of Psychology.*

[2] Dr. Tchelpănov presents this thesis excellently. "Color sensations are primarily of an objective order. Were they as innately subjective as emotions, neither muscular feelings, nor any other additional moments would be able to give color objectivity, exactly as no association is able to invest our emotion with a feeling of something objective."

conditions of color perception, must lead to a granting to space primal quality.

Had the primal trait of irreducibility to simpler substances its source in the empiric subject only, it could have no other meaning than being inborn. Given such primacy, the spatial trait must be so considered.

But in explaining the rise of spatial perception in personal experience, in making clear its inherence in the empiric subject, one has to presuppose the formation of the space concept during the genesis of the species. Which does away in any case with the proposition that space quality can be reduced to simpler components. With the genetists straining to show that the sensation of space or expanse can not be primal, one comes – again and again – to the conclusion that the irreducibility of the sensation of space has an epistemological meaning, not a psychophysiological one. That is, spatial order is a pre-ordained form of perception or a law of perception. Its spring is the trancendental subject which fathers the pre-conditions for the experience as a whole.

Here is a solution for the above antinomies – among them, the antinomy related to depth. The latter is pitting an excellent proof of the primal quality of depth given by James and Stumpf against the equally potent critique of Berkeley and Lipps based on the impossibility for our senses to perceive empty space.

As a sense datum, i.e. as a function of the empirical subject with its psyche and optical apparatus, the fact of space could amount to only a colored flat continuum. As a function of the transcendental subject conditioning the form of perception and of the very organising of sense data, the spatial quality lends empty space a certain meaning.

But only a merger of the two principles could generate the existence of depth.

An inquiry into the meaning of geometrical generalizations, an issue of the definition of space as order of perception – in other words, as law of organization of experience, such inquiry sets the absolute core of geometry as an ensemble of traits baring which space in its higher abstraction loses its epistemological meaning: that of the general form or pattern of experience.

III

In a mathematical generalization its particular point is that it is in no need of a preliminary existence of the ideas united in a parent concept. Mathematical abstraction creates the variations out of itself, and, together with the initial concept (basis for generalization), ranges all within a new parent idea.

Variations are created this way. One of the features of the subject for a generalizing, is considered a variable; and concepts derived from variations in that feature, bring new and wider abstractions.[1] Such a trait becomes basis for a mathematical generalization.

Metageometrical abstractions' basis are either measurable, metric traits of geometrical symbols, or projective. Generalization based on the former yields the parent idea of space continuum of a constant curvature and n dimensions where every dimension is an imaginary variable, in a private case – a concrete one. This variable is subject to any and all mutations from $-\infty$ to $+\infty$.

Abstraction or generalization based on projective traits yields the parent idea of space as a group of projective transformations.

METRIC TRAITS

α) *Curvature.* In mathematical respect, and given the same number of dimensions, the measure of curvature appears as the sole metric evidence of difference between spaces of a constant curvature. If two space *continua*, two surfaces, for example, have the same curvature – they are identical. Their nature is described in the same geometrical terms.

Mathematical generalization goes even farther than that.

In a geometrical system still more general and formal, it substitutes parity by symmetry for parity by superposition.[2] Considers, for example, identical the same symmetrical figures on a spheric surface which can not be superimposed through a motion that does not also change their relative position.

As long as these abstractions are taken as a scientific utility,

---

[1] A case in such abstraction is, for example, mathematical induction which mathematicians erroneously call complete induction: a conclusion from n cases to the $n + 1$ case. (Wundt's "*Deduction nach exacter Analogie.*" See Wundt, *Logik*, II Band).

[2] Beltrami: "Theoria fondamentale degli spazii di curvatura costante" (*Annali de Matematica pura ed applicata*, serie II, Vo . II.).

they are purposeful and logically faultless. But the moment geometricians use them to sustain an argument of epistemological nature, investing these abstractions with a broader meaning and a higher generic sense, such generalizations start gathering fallacies.

As for figures identical through symmetry, it is evident that they can be so considered only in a mathematical sense: by no means psychologically or epistemologically. There certainly is an "imponderable" that causes such figures to differ – in spite of the complete coinciding of metric traits. That moment or "imponderable" can not be defined mathematically.

Unrolling figures present the same case.

In a geometrical sense there is no difference between figures sketched on separate surfaces but identical upon unrolling: all mathematical traits, curvature among them, fall in correctly. The said figures do not present, however, identical mental facts. As such, these designs are characterized precisely by the general configuration of points, the manner of curving and of position of the entire figure: none of which is really identical.

The following example uncovers the distinction between the geometrical and the psychological or epistemological meaning of the abstractions in question.

A living being of one dimension with a rigid stem of a very small but finite length and zero curvature, dwelling on a cylindric surface, would notice that it can move in directions only parallel to itself and the cylinder's forming radius. Such a creature would evolve a geometry devoid of the axiom of free mobility, a geometry totally opposed to one that should be conceived by the above hypotetic being according to the geometrical definition of a cylindric surface.

Does this mean, however, that such a surface is a space of changing curvature? Taking such surface for basis, could one insist that spaces with changing curvature are imaginable?

We shall see later that such imaginary continuum does not make *gnosticological* sense. Mathematically, a cylindric surface is defined as one of a constant, namely, of zero curvature.

The above contradiction comes from the fact that the mathematical generalization is artificial and uncertain as to its epistemological aspect.

Of the two radii of curvature of the main cuts or sections in a cylindric surface, one presents a certain positive quantity k, the other is of an infinite length. The general curvature of the surface at each point is expressed therefore as $\frac{1}{k} \cdot \frac{1}{\infty} = 0$. Thus a cylindric surface is coupled in the same abstraction with a flat one, which is indeed mathematically correct.

But in essence, the multiplier $\frac{1}{\infty} = 0$ which reduces the measure of curvature to zero, is precisely the factor infusing an imaginary moment into the formulation of cylindric surfaces. Bringing this uncertainty to a clear general meaning, one perceives that a cylindric surface is one of zero curvature in the mathematical sense, but does not possess this trait in a strict gnosticological sense: when thought of as a feasible form of outer experience or dwelling space.

Our one-dimensional creature of a tiny rigid stem would feel the zero curvature only when moving along paths parallel to itself or to the cylinder's forming radius. In other movements it should experience a positive curvature if it could get into such positions. Of course, I do not wish to state in so many words that a living being can actually experience one or another curvature; it should simply feel a sensation of otherness that must come with a change of space's curvature.

But even if the mathematical abstraction were correct in the above sense, it remains subject to still another limitation.

Driven to its logical confines, our geometric generalization should have created the broadest abstraction – that of space with an ever changing, limitlessly variable curvature.

But without any *mathematical* necessity, the metageometric generalization halts on the threshold of the above vision, limiting itself to the idea of space with constant curvature. This is caused by our willing adoption of the premiss of dimension that presupposes free mobility and continuity of space.

The inner geometric relations are expressed in terms of curvature and therefore allow measurement when the space is continuous and the terms of measurement do not depend on the selection of coordinates. Then functions defining locations have derivatives by coordinates; and space is thus endowed with free mobility.[1]

These premisses are, of course, of no mathematical necessity, for spaces of changing curvature are feasible (see the first chapter of this treatise). The above premisses are imperative only in gnosticological respect. Embodying, in form of the axiom of free mobility, the absolute part of geometry, they show that such part can have no other meaning save the epistemological. (see chapter II here).

Q

Thus, the premiss of free mobility limits the generalization based on curvature.

Russell sees the necessity for this premiss as a dual one. Firstly, it conditions the very possibility of dimension; then, it expresses the basic nature of "outsideness" (Russell thus generalizes the idea of space). "Outsideness" affirms the relativity of all positions in space.[2]

But the first Russell consideration can not, in my opinion, be in itself the ground for the axiom's indispensability: even if for the sole reason that spaces lacking free mobility are feasible. Russell's point means no more than this: free mobility is a pre-condition for the possibility of geometry, not space.

Nor does Russell's second point establish the necessity for the axiom with sufficient strength. Deducting at times all the traits of his form of "outsideness" from mere dialectics, Russell's very idea of "form of outsideness" sketches but feebly the psychological or epistemological principle that would demand relativity of position.

What would happen if space were not homogeneous, Russell describes clearly:

"Space would be then no more passive. It would exert definite effect on things, and we would have to accomodate ourselves to the notion of points marked in empty space, these points being indicated not by the bodies hemmed in by them, but by the

---

[1] Helmholtz: *Wissenschaftliche Abhandlungen*, Vol. II. See also Riemann's *Gesammelte Werke*, p. 280 etc.

[2] Bertrand Russell: *Essai sur les fondements de la géometrie*.

effects of space on all the bodies that could from time to time occupy those locations.

But this lack of homogeneity and of passivity is absurd. For space is a form of "outsidedness;" can not admit positions others than relative and non absolute; must be completely homogeneous from one end to the other." [1]

But why should not positions in space shape objects? Where lies the absurdity of such a proposal?

Russell proves efficiently that wishing to measure something in a space deprived of free mobility, we are bound to tumble into a *regressus ad infinitum.* He describes very well, indeed, the geometric absurdities derived from the absence of free mobility. But in the explaining of the reason for such absurdities he only skirts the source of necessity for homogeneity of space; does not get to the right point.

Space devoid of relativity of positions would be senseless as an order of perception or form of experience (see my chapter II in this essay). The absence of homogeneity of space, its changing or absolutely variable curvature would mean that the order of perception may be basically indefinite, may even change at any given moment.

Such a proposition is psychological and epistemological nonsense. As a pure form of experience, the law of perception is firstly the primal generator of every change. Secondly, not investing our perception with any material contents, the order for our grasping of sense-data or perceptible things is uniformly applicable to the latter, binds them in a uniform manner. *From this source only* flows the necessity for free mobility and the demand by the later for continuity of space with its mathematical corollary: the possibility of differentiating functions of position.

The above argument brings us to the following conclusion.

The gnosticological limit of generalization based on curvature shows that, in the question of the *a priori* essence of axioms and of the feasibility of geometrical systems others than that of our common experience, the deciding factor is not the mathematical but the gnosticological possibility of such novel systems; the definite epistemological sense they may make.

---

[1] Russell: *Ibidem.*

Because of this, a purely mathematical abstraction not presenting an idea *gnosticologically* wider, can not by itself serve as a basis for the argument that if non-euclidean systems of space relations are feasible, our common geometrical ideas have no necessity.

This, as we shall see, pertains also to generalizations based on traits other than curvature.

β) *Dimensions.* The expanding of the idea of space based on this trait brings forth the parental form of a *n* ways' continuum where dimensions are complex variables. In such a space the position is defined by the function $\psi(x_1, x_2 \ldots x_{n-1}, x_n)$, with entities that may freely vary from $+\infty$ to $-\infty$.

An extreme abstraction as such has only analytical value, and, for our purposes it is sufficient to look into a less general idea of a continuum made up of concrete dimensions. This being a result of reasoning from *n* cases to $n + 1$, one must again read correctly the epistemological meaning of such a generalization.

Analysis of space perceptions has already shown that spatial quality is indivisible and indissolvable. A direct sequence of this: the idea of dimension is an abstraction. In relation to the essence of space, dimensions play the same role as the system of coordinates in defining geometric positions.

Dimensions of common use can possibly be discarded for other dimensions, different in number and character; but the number and nature of dimensions in *our* use are the ones that firmly set relationships in *our* space.

Therefore, to some extent, those mathematicians are right who show that were we to take as key element of a continuum not the point but a form dependent on an indefinite number of parameters, space could be imagined as a continuum of any number of dimensions (Plukker). [1]

But the fact that dimensions are moments generated by mere reflexion defining relationships, has no bearing on the epistemological meaning of dimensions. They still can affirm, not directly yet powerfully – the traits of space that do not depend

[1] See Felix Klein: *Betrachtungen über neueste geometrische Forschungen*, S 5.

on dimensions: exactly as the invariant expressing the distance between two points $\Omega\ (x', y', z'; x'', y'', z'')$ does not depend on the choice of co-ordinates.

Having applied the above to the interrelation of two- and three-dimensional space, it is easy to see that stereometric space can not be a generalization or broader generic idea in relation to surface.

In the first place, the idea of surface is pure abstraction. In the second, the mathematical definition of surface as two-dimensional space does not hold water in psychological and epistemological respect. Therefore, in their deeper sense, surface and space do not stand in any such relation as the continuum of two to that of three dimensions.

Having defined the idea of geodesic line on a sphere, conditions of its motion on such surface and the concept of a spheric angle as part of sphere hemmed in by two geodesic lines, etc., it should be possible to evolve spheric geometry without using a single stereometric symbol. Mathematicians hold this as proof that the sphere is a two dimensional space.

But it may be asked if the perception of sphere, no matter how its traits and definition have been evolved mathematically, if the very perception of sphere would, in the course of a mathematical unfolding, lose that indispensable quality of a form of perception, the quality of an indivisible continuum; if the perception of sphere evolved mathematically would then make epistemological sense.

Stumpf's discourse shows, in my opinion, that a flat or curved surface is inavoidably related to three-dimensional space. Therefore, while geometry of any kind of surface could be construed independently from the third dimension, the psychology of space can not be thus evolved.

In epistemological respect, surface can mean only-an object of perception; and as such, it is clothed in an indivisible space form. As a self-contained form of perception it has no reality. The stereometrical moment does not simply add itself to the quality of a continuum. What makes of the expanse or continuum a form of perception, that is, live space, is precisely that stereometrical moment.

It has been seen earlier that free mobility is a mark of any order of perception. As there exist surfaces with absolute positions, such as the one-sheeted hyperboloid, surface does not necessarily possess free mobility. This serves again as an indirect proof of the proposition that the law of perception translates itself into stereometrical aspect of expanse.

Thus, epistemologically surface can not be considered a frame of space. The generalized abstraction of an n-dimensional continuum loses gnosticological sense in case of two dimensions. And as our reasoning could apply to one-dimensional continuum (line) and to zero-dimensional (point), these two forms, too, lose the meaning of space.

Such facts alone would make amply clear that a generalization based on mathematical induction, reasoning from n cases to $n + 1$, – has no power in a gnosticological problem. Keeping the basis of generalization in its mathematical unravelling logically safe does not guarantee correct presentation and reasoning concerning the psychological or epistemological order of the same sense-data when subjected to even the highest abstraction.

But the continuum idea that establishes the equal possibility and equal validity for all cases, with number of dimensions varying from zero to infinity, loses philosophical meaning not only in the case of one, two or zero dimensions. The case of an infinite number of dimensions is equally short of gnosticological sense.

Again Russell bases the necessity for a finite number of dimensions on such a necessity being *conditio sine qua non* for any setting of space relations.[1] Which proves, of course, that a finite number of dimensions is a postulate for geometry, not experience: it does not in itself prove the *a priori* nature of the axiom. The necessity for such a postulate is by no means a of mathematical order: it lies in the truth that the law or order of perception can not be indefinite.

Thus, there dwells a limit to abstraction following the unravelling and broadening of the dimension-concept. The source of this limit curtails the reach of a purely mathematical necessity.

If, however, some cases of our abstract space-continuum – that of $n = 0$; $n = 1$; $n = 2$; $n = \infty$ have no concrete spatial

---

[1] Russell, *Essai*, § 158.

meaning; are gnosticologically in another class than the case of three dimensions, – then one can not be sure that there are still other *continua* with a finite number of dimensions, other than three which have gnosticological meaning. The three dimensions *may* be a necessity. In any case, it is not the mathematical feasibility of space with another number of dimensions than ours, that refutes such necessity.

Therefore, I can not acceed to a view that considers the number three in the axiom of dimensions just a sense-datum.

The third dimension is a quality of space separable from it in speculation only. Were the essence of dimensions to be seen in their affirming of the nature of space but indirectly, then dimensions would appear to be indicators or symptoms of the order of perception. By this order they are bound as coordinates by an invariant. It follows that the necessity for three dimensions flows from the same principle as the necessity for a finite number of dimensions: from the inalterable order of perception.

The law of perception taking on a form of space apparently binds its phenomena, the sense-data, into a three dimensional pattern.

γ) *Distance*; The abstraction broadening the concept of space along the trait of distance, is based on the generalizing of the idea of straight line as geodesic.

Here, too, the generalizing leads, by way of mathematical induction, from two-dimensional space to one of many dimensions. Into the distance between two points any number of dimensions is "wedged" in. In this way the equation for the element of distance in a multi-dimensional space is arrived at:

$$ds = R\sqrt{d_x^2 + d_{x'}^2 + \ldots + d_{x_n}^2}$$

where R is a constant.

Geodesic lines – shortest distances between two points-bring to zero the variation of an integral with subintegral function below.

$$\delta \int \sqrt{d_x^2 + d_{x'}^2 + d_{x'}^2 + \ldots + d_{x_n}^2} = 0.$$

Unfolded, this formula defines the nature of a multidimensional continuum of constant curvature.[1] It is clear then that general-

[1] Beltrami: *Theoria fondamentale.*

izing based on distance mounts again from a straight line on a surface to a straight line in space.

Q

From the preceeding section one has gathered that only mathematically can a two-dimentional continuum be deemed a form of space. Needing for its own perception some three-dimentional points, the two-dimensional continuum has no epistemological equality with the former. Because its subject is only the straight line in two-dimentional space, the discussed generalization cannot stand scrutiny – if for the sole reason that it is built as a mathematical induction.

But aside from that, the idea of a straight line or distance, being a function (or "derivative") of the order of perception, must be a predicate to perception, to begin with; and secondly, derive its meaning from the acting order of perception: the euclidean space order.

The first point comes clearly from a critique of the evolutionists who view the straight line as a hypothetic idea perfected by experience.

This explanation was seen by Dr. Tchelpănov as a vicious circle. "In the evolutionists' theory, in order to reach the abstraction, we must first discard all that mars the perfection of the idea of a straight line; and therefore possess the concept of an ideal straight line beforehand."

Wundt proves almost the same: "*Der Gedanke der lagenbestimmenden Verbindung zweier Punkte ein Postulat unseres Denken ist, keine wirkliche Vorstellung.*" [1]

The idea of straight line is created *a priori*, for it is a function of the order of perception. But in this case, the straight line concept can not embrace traits conflicting not only with the law of perception in general, but even with the actual machinery of perception. Yet the mathematical idea of space generalizing the straight line as geodesic, leads us precisely to this contradiction.

To start with, abstraction brings us to spheric space where one finds an infinite number of straight lines between polar points.

---

[1] "The idea of relation between two points that defines their position is a postulate of our thinking, not an actual perception." Wundt; *Logik*, Volume II.

Which is upsetting in gnosticological respect. The relative position of two points which should be defined by distance only, would then admit an infinite number of relationships. Which would mean, in turn, that the order of perception can be indefinite.

Trying to unite the seemingly unlinkable – the highest mathematical abstraction with gnosticological sense – Russell suffers pains from the fact that spheric space admits more than one shortest distance between some pairs of points; and this – in spite of his axiom of metric geometry. But his explanation (see *Essai*, p. 169, etc.) is so feeble that it is not worth quoting. As to Russell's reasons for the axiom of distance, I must again counter them with my remarks in chapter III, *a* and *b*.)

The second of the above mentionned conflicts is the finite straight line of elliptic space and, consequently, the finiteness of the space itself.

Epistemologically this, too, makes no sense. In its pure action the order of perception allows no restrictions; it must create positions in no way exhaustible and always related.

In euclidean space *all* positions are related, privately or as a whole. In elliptic space, however, positions being relative privately, are absolute as a whole, because the position of a point on a straight line which is both boundless and finite, cannot be deemed relative to another in the true sense.

Thus generalization developed from the idea of distance is also limited by a gnosticological necessity: the predicate affirming that two points have only one relationship – their distance. A predicate that has no *mathematical* necessity.

PROJECTIVE TRAITS

Generalization or abstraction following these characteristics is based on the principle of projective geometry according to which geometrical objects are identical when they are obtained by mutual projection that does not change their nature: i.e., does not alter qualities bearing solely on geometry of position.

As for this abstraction, one has to repeat everything said earlier. One could show again that projective junctions are objects identical in a mathematical sense only; that projective correlation sustains but an artificial generalization epistemologically muddled. They can not support argument in a purely epistemological problem.

Abstraction based on projective traits also has its limits.

The result of such generalization is the parent idea of a group of transfigurations which could be better called *collineations*. These possess the following trait: a combination of any two collineal forms results in a new transfiguration itself becoming a member of the group. Such a concept embraces all groups of lineal variations, among them – the groups of projective junctions linked by an invariant: the main group, for example, which does alter distances, that is, relations to imaginary circling points at infinite distance.[1]

Mathematically the latter is a private case; gnosticologically – a general. For, as it was shown earlier (in chapter III γ), the existence of the above mentioned invariant is a condition for the definiteness of the order of perception, in fact, for its existence. Without the possibility of translineations that do not alter distances, space loses its meaning of a form of perception.

Thus the concept of space as a group, even if based on a broadly general idea of distance as relation to a certain conic section – which does not in itself bear gnosticological scrutiny – even such an abstraction has limited use.

The mark of this limitation is seen not only in the above specific idea of distance, but also, in the very definition of space as a group.

Mathematicians persist in saying: "Space is a group." This is wrong because a space of variable curvature is feasible: one where a fusion of any two translineations does not always result in a new member of the group. Which means that mathematicians should correct themselves and say: "Space *must* be a group."

As, however such a proposal flows from no mathematical necessity, there ought to be a gnosticological demand, and namely this. Being but a group, space means an order of perception that binds our sense data in a uniform way.

The above discussed abstraction is the cherished argument of mathematicians in favor of equality or equal applicability of the various geometric systems.

---

[1] See Felix Klein: "*Nichteuclidische Geometrie*" (*Mathematische Annalen*, Band VI).

Poincaré asks, what do such hypotheses mean: existence of a group of translocations that do not alter distance? "Are they sense – data or judgements, analytical or synthetic *a priori*?" Each of the three questions he answers in the negative.

"Were all these hypotheses facts of experience and observation, geometry would be subject to everlasting revision, not an exact science. Were they synthetic a priori, and even more, were they analytical judgements, it would be impossible to forgo them, and nothing could be built on such a negation."[1]

However, *one can refute* these hypotheses – and build on such a negation a system logically and analytically feasible. That is, one can "materialize" the second member of the disjunction, the first of which is the positive proposal presented in chapter I α and β.

Furthermore, these hypotheses, in Poincaré's terms, or axioms – in ours, define the nature of the inalterable order of experience. They *can*, therefore, be facts of experience and observation. And yet, geometry does not have to be subject to revision; not by this token, anyway.[2]

Thus the seemingly powerful Poincaré disjunction does not prove at all what he meant to show: that the choice of a geometrical system is based solely on the convenience of applying it to space in our use.

IV

The abstraction of the concept of space is a sort of logical multiplication; for out of all terms qualifying the class, it segregates but the most general, with the absolute core of geometry as result.

Mathematical logic shows that logical multiplication – just as the other direct act, arithmetical multiplication – is always achievable. And the result is never zero, in this case.

But the multiplication of all possible terms inherent in a class of mathematical space, does result in a zero, for no trait common to *all* feasible geometries exists.

Free mobility and the existence of a group of exchangeable

[1] Henri Poincaré: *Sur les hypothèses fondamentales de la géometrie.*

[2] The same line of reasoning in relation to arithmetic is seen in my essay "The Roots of Arithmetic," in this book.

collineations has no place in space of variable curvature. The inherent characteristic of the "shortest distance" between two points is invalid in space of positive curvature. And the general traits of an n-dimensional continuum lose necessity when $n = \infty$.

But if the multiplication of terms defining a class of mathematical space is impossible or senseless, such class is impossible. Which results from the situation that there are members in this class that exclude each other.

This takes us back to the opening of the present essay and explains why the argument of feasibility has no power.

"If space relations other than ones in our common use exist, then ours have no necessity," this proposition includes a variable term: "space." And members of this class excluding each other, we are facing the fact that the above thesis contains but a logical disjunction which, in essence, reads thus:

"As the two are equally possible in mathematical respect – a space with such and such inherent traits, but also, a space excluding such qualities, it is clear that the geometric relations in our common use have no necessity."

The argument of pure feasibility is then a proposition with a variable term for space. The thesis would have a precise meaning only if a constant term is substituded for the variable. As the concept of mathematical space forced by claiming logical equality for all its variants, is the very thing that takes any vestige of definiteness out from the argument of pure feasibility, we are thus bound to substitute for terms defining space other than mathematical.

I have shown that the absolute core of geometry obtained through abstraction (by logical multiplication of feasible space concepts), that such absolute part exists only by the grace of epistemological necessity. No matter which way mathematical abstraction is led, one always stumbles on a trait indispensable epistemologically. This trait is invariably the one that limits generalization.

Beyond such limit the mathematical concept loses general sense, and the mathematical abstraction – its proving power. Generalization based on free mobility is deprived of strength by the necessity for constant curvature. The abstraction generalizing the trait of dimension is enfeebled by the demand for a finite number of dimensions. The abstraction based on distance is viciated by the postulate of only one shortest between two points. The widening according to the projective traits runs into the necessity for existence of an invariant of projective translineations.

Once the mathematical generalization is limited by epistemological demands, we are forced, in our standing argument of feasibility, to substitute for the variable – not the mathematical but the gnosticological concept of space, that is, the order of perception.

It is the possibility of an order of perception other than ours, that speaks against the necessity of euclidean concepts: not the feasibility of other geometric relations.

Philosophy could therefore cut into the problem of the *a priori* nature of euclidean axioms by the sole means of psychology and epistemology. But of course, the power of analytical and geometrical illustration, the clear shaping of problems and mighty analysis that mathematics manage to perfection, can supply epistemology with details that might slip away from an abstract critique of reason.

In any case, however, it is idle to think that the epistemological question as to the *a priori* essence of the euclidean concepts is naturally resolved by a mathematical abstraction. ,

The mathematicians should address to themselves what they so often tell others: the words of Beltrami.

"Se la demonstrazione è stata concepita in vista di una determinata categoria di enti, senza che in essa sieno stata effetivamente introdotto quelle determinazioni che individuano la categoria stessa in confronto di una categoria piu estesa, è chiaro che le consequenze della dimonstrazione acquistana una generalità pui grande di quella che si cercava."

"If proving statements were made in relation to a certain definite category of things without stipulating the terms that delimit this category from a broader one, it is clear that the conclusions of the above proofs will acquire a generality wider than warranted." [1]

---

[1] Beltrami: "Saggio di interpretatione della geometria non-euclidea" (*Giornale di mathemathiche,* 1868, vol. VI).

The Philosophical Value of Science

I

Aware of the fact that the "outer" world is but an image or a figment of imagination, the man considers this world of ours a phase of his own spiritual experience; a phase perhaps more stable than the others.

Psychology grants that the "outer" world is a sum of organized sensations, and admits – at times but obliquely – still another, a real world outside our consciousness; a sum of "things" of which, through a film of our impressions, we obtain vague images.

Such view does not, however, differ from that of nature science; the latter is only more pointed. Sciences with a psychological basis – anthropology, theory of language or history – accept the above viewpoint.

Embracing spiritualist or psychophysiological schools which admit the interaction of spiritual phenomena and physical, psychology yet refuses to explain the link between physical nature and our spiritual states. But in the general philosophy of science, that is, in the system of co-ordinated higher conclusions of science amounting to a theory of the world of objects or empirical cosmology, the view of spiritual phenomena must revert to the starting point of nature science and base itself on psychophysiology.

Yet in explaining cosmos, philosophy of science touches upon only one aspect of the universe, its empirical validity: as a philosophical approach, a view limited and relative; one to be filtered through a broader viewpoint – looking at Being from every angle.

The latter approach evolves *its own* gnosticology, truly philosophical and absolute. For it flows from the one certain truth, immediate in our awareness: the world is *my* concept.

We thus come to the idea that the test of value for every philosophy of Being resides in the criteria of transcendental idealism. Such viewpoint *can be* wedded to pure materialism – *in science.*

Another source of the above thesis is the certainty of the basic condition for the very life of science: the immutability of the laws of nature. This assurance is inherent in the awareness of the transcendental ideality of the world of objects, of that world's *functional* dependence on the forms of perception.

Broader than the theory of knowledge based on science, the above viewpoint serves as arbiter for the philosophy of science. This should be made clear to anyone caught in a vicious circle: meditating on knowledge and considering the propositions of science only.

The world is a concept, a figment of imagination. Whose imagination?

Nature science and psychology say that is an image within *my* awareness, in itself a function of my brain. Which means that action of the brain where sensations are gathered and handled, is something the world is conditioned on. The brain is thus both, the instrument and the object of imagination: the vicious circle just mentioned.

If the world is a sum of images, it is imperative that there exists an imagining center, or a being playing this role for all feasible objects, without amounting itself to but a figment of fancy.

Such an instrument of imagination and receptacle or birthplace of images must then be of necessity transcendental. For only in this case is our unshakable certainty of *"I am"* possible. Only in this case can the subject be the holder of the transcendental forms of Being.

This vehicle or marshalling center for images making up the world of objects should be properly called transcendental subject. Its nature and limits are unknown to us. Its meaning lies only in the functional correlation with the world of outer events. In this it partly "covers" the empirical subject.

For even Kant does not make it sufficiently clear *who* or *what precisely* is the "keeper" of this world, one seemingly outside our body or outside our receiving self. Kant had merely established that whoever this "keeper" or "receiver" of sensations from outer objects is, the dependence of the outer world on such a receiver must be functional.The frame of the outer world, its very forms and conditions of existence, must be inherent in the above "receiver" or subject.

Having thus formulated my viewpoint, I shall turn to my task: mainly – to the question of what do sciences contribute to two formal disciplines: the empirical or scientific theory of knowledge, satellite of philosophy of science; and to philosophical gnosticology, companion of metaphysics.

The two theories of knowledge do not exclude each other, just as the empirical aspect of the world does not exclude its transcendental, ideal nature.

One should be aware, however, that supreme truth resides only in the philosophical view of knowledge. The latter's empiric aspect shows its conditional and relative nature in the earlier discussed vicious circle.

We see it take on the following form. The nervous system of an animal is both, the product and the parent of the environment or outer world. Again, only the prevision of a transcendental subject where forms and conditions of existence for the outer world repose, in which the latter's history unfolds, only such proposition can resolve the above vicious circle.

II

Our "knowledge" or ability to co-ordinate our sensations touched off by the outer world, has grown out of animal life's necessities.

Leaving out the question of how nervous life or energy originated, and whether it even could "have begun" to exist, let us imagine the history of some sort of protozoan with but most elementary nerve functions living in some physical milieu.

The activity of such a protozoan strives to preserve its personal and generic existence which demands knowledge of the milieu. If through some occasion or complex cause, its sensorium has developed a change that permits it to grasp still better the nature or state of its environs, that is, allows for a better equipped struggle for existence, then the change giving the protoplast such an advantage, fortified by natural selection, will give it a new organ of reconnoitering – through heredity.

Nerve life with its summit, consciousness, grows with the refining of the physical organization, or with the increasing difference between the personal physique and the milieu.

Perhaps, the very awakening of nervous life in protozoa with rudiments or a semblance of nerve formation, is conditionned on the above disparity.

But let us turn to what – to an animated being – seems a symptom of the state in a milieu: the signal or the sense datum.

It is self understood that sensations being born as reaction to an outside world, the organ of knowledge of the latter should be the sense of touch, this or another form of it.[1]

Even a superficial probing makes us see that the so called outer senses reduce themselves to the sense of touch, either immediate or indirect.

With the nerve ends of the retina the eye is touching not indeed the object itself, but the rays deflected from it. And through the angle of sight and muscular effort of accomodation and of converging the axes, the form, size of and distance from the objects are gaged.

Color reactions of the nerve ends to waves of ether varying in length, are still another variety of sensations of touch.[2]

---

[1] The discussion at this point is, indeed, not strictly scientific or even only specific, but philosophical and general. Offering but a scheme of happenings and a very general interpretation, this passage is bound to look as an elementary approximation.

[2] According to the Jung-Helmholtz theory, some fundamental colors are tied to corresponding nerve-ends of our vision's apparatus. See Helmholtz: *Newest Achievement in the Theory of Vision.*

It is easy to show that in their outer aspect, that of perception, all other senses, too, reduce themselves to the sense of touch. As to the inner aspect of sense-data and their spiritual lining, so to speak, such a generalization would be crude and faulty.

Turning to the history of senses, difficult as it is to establish the order of their appearance, we still find it confirmed that the fundamental instrumentality of sensation is the sense of touch; and that the rest of the forms of sensation have sprung up because of incidental changes in the existing organs retained through natural selection as an advantage in the struggle for existence and transmitted all the way down the specie through heredity.

But if in their prospective nature all the senses actually amount to a variety of touch, the inner aspect of sense-data, or their spiritual lining – so to speak – could not be summed up in this sweeping way. As to its spirituality, every form of sensation is *sui generis*: does not lend itself to so primitive an interpretation.

The refinement of animal sensoria starts with accidental changes transmitted down the specie.

Our outer senses gradually manage to gather minuscule, complex and imperceptible changes of milieu.

In sensations of light, the changing length of various waves of ether brings us a variety of colors; the length of air waves – various sounds. Not always can more complex sense – data be dismembered into component parts by our organs of perception.

Sight, for example, is *no more* capable of dissolving fundamental and transitory colors into components or supplementary shades. In aural sense-data only a very acute hearing can grasp the component sounds of a consonance, still less - of a dissonance. But timbre which is the sense-result of vibration of a note *with* its overtones, can not be broken up into components by even an ear of incredible subtlety.

Of course, only the sensation is indissoluble. Physically, the most complicated sound wave can be dissolved by a resonator, and the light wave – by a prism. Thus our nature strives to point to us a change in the outer milieu by a simple new sign necessary for reconnoitering.

This is why primitive, fundamental sensations arrive first. Alterations in the organs of perception bring corresponding and subtle variations of the sense-data themselves, such as timbre and harmony in the sound sensations, transitional shades in colors, and so on.

Where this formerly composite sense datum turned into a unitary and indissoluble one, it is coupled, as an easy and well known means of reading the environment, with a sense of pleasure. Where, however, the sense datum is still composite and more or less irreducible to recognizable ingredients, like dissonances or some complex odors, its being a difficult and unfamiliar way of understanding the environment leaves in the perceiving subject anxiety and displeasure.

With the man, all sensations are an issue of the upper layer of nerve life and thought that changes and varies the meaning of sense data. Besides, owing to the peculiarities of the man's world, the law of natural selection and accommodation to nature has a weakened effect.

In the animals' psychology feelings and sensations are unmixed; their related signs pointing to state of milieu are clear and automatically right.

III

Our thesis then is that with the development of the organs of our senses only specific sensations change (or new ones come in); that given an unchangeable environment, the instrumentality of senses amounts to the same medium of touch. The former only passes through a higher refining process; comes to perceive more and more minute details of the milieu. Substantially, all specific sense organs have unfolded as a variation of touch.

This main proposition with all its details is sustained by major facts and conclusions of embriology and physiological optics.

According to the fundamental thesis of embriology the individual form passes through the same stages of development as the species: onthology repeats the philogeny. It would be enough to view the philogenic unfoldment alone.

With the most primitive animated beings whose body presents a simple skin-bag; or with those complex in structure where all the three primordial skins, outer and inner develop into body organs, their tools of perception remain essentially instruments of touch, either direct or in a differentiated form.

Still another proposition of importance that sustains my analysis, is the law of physiology of vision discovered by Johann Müeller: the so called law of specific energy of the organs of our senses. This law, by the way, could be easily forseen through a philosophical *a priori* reasoning.

According to Müller's law, the sensation depends solely on the organ being excited, not on the cause of the excitement or the nature of the excitant.

The wave of ether may evoke a sensation of touch: the infra-red part of the spectre may produce the feeling of warmth.

Every organ that conveys phenomena pertaining to it, transmits generally any changes of milieu, just as the ordinary sense of touch does.

Yet, as every state of the environement could, exciting any organ of our senses, evoke a specific sensation and mislead the animal, the natural selection has sheltered the specific organs in such a manner that only definite and corresponding excitants can produce the effect. Thus, the eye sitting deep in its socket, protected by the bone frame and eyelids, is hardly accessible to other agents save the light waves.

Thus the vicinity that excited the organs of our senses is constant, while the sensations corresponding to the states of the milieu change in accordance with the organ that is being excited. And as the sense of touch is fundamental, essential and relatively true – that is, known to all animated beings and having the same meaning for all, the nature of the outside world must be, it seems, purely mechanical.

It should be imperative then to reduce some of the physical traits of the outer milieu seemingly non-mechanical, yet grasped by the sense of touch (heat, for example) – to mechanical qualities. This shows once more that any means of perceiving our outer vicinity, no matter which specific sensations mark such perception, are a form of touch, and that, on the other hand, our outer milieu is uniform and possesses only mechanical traits.

Aside from the fact that all the variety of sensations lead to one agent (all colors – to waves of ether of different length), the excitants of diverse sensations are also traced to one source. As the experiments of Herz have shown, electrical and light phenomena are the result of the same kind of agent, the waves of ether which become refracted, reflected, interfered with and polarised.

These agents are material and possess definite mechanical qualities: hardness, a known co-efficient of elasticity, in Maxwell's formula, and evidently, a certain active energy in their particles. [1] Heat, too, is reduced to latent movements of physical particles.

It is clear then that any sensation is an inner sign accompanying the touching of particles of matter in movement, of matter having purely mechanical characteristics.

Chemical qualities of matter, too, are reduced by new research to multiple motion within the molecule. And so, the general viewpoint of physics is valid also for the world of chemical phenomena.

The sequel to the above discussion comes to this.

In order to understand the outer world, appraise its relative truth, one has to think of this world as it would present itself to us were we to receive no other sensations than those of sheer touch, the sensations of a protozoan.

Such a world is that of matter marked by sheer mechanical qualities, in Kant's words – "the mobile in space."

IV

Here is the frame of a consistent scientific theory of knowledge, the facts and considerations lying at its base.

How much can such a gnosticology, how much can its facts be ground for philosophy?

In any case, no matter what the nature of the outer world in which the subject – animal may live according to scientific epistemology, its reality is only relative. For that outer milieu, its potential and history, are born by the transcendental subject. But let us see regarding

[1] Clarke Maxwell: *Ether*, Addresses and Essays.

which traits of the environment and to what extent it is possible to claim even such kind of reality.

Whether new kinds of perception appeared, or the old ones never altered, all phenomena of the outer world would have the same degree of relative reality. They would stay uniform as to their gnostico-logical meaning.

But first, beside the purely tactile ones there emerge other kind of signs and symbols which have a meaning to the animated beings endowed with special organs. The more kinds of perception appear with correspondingly differentiated sense-organs, the narrower grows the circle of animated beings for whom the emerging signs are of actual effectiveness. Which means that their relative reality tends to diminish.

Secondly, the variety of shades in the sensation itself, such as timbre, composite collor or odor, reach the animal directly as simple, indissoluble signs. The latter are related, however, to two or several simultaneous changes of the physical milieu: sounding of the basic note with its harmonics, mixing of color shades, and so on.

Primary (tactile) signs stand in simple relationship to the changes of environment. Mathematically, the present is a still undefined but simple function of these changes. This is why primary signs hold the first degree of relative verity, and their corresponding phenomena – the first degree of relative reality.

With the emerging, still later, of secondary signs, it is different. They do not correspond to the changes in the milieu as directly, nor as obviously. I have already shown that the new, complex signs are alterations of the primary ones. The new signs emerge because of the alterations in the primary sense organs.[1]

Mathematically then, new signs or symbols are to be functions of the primary signs, that is, a complex function of the physical milieu. This is why these signs enjoy but the second degree of relative verity, and their corresponding outer phenomena – only the second degree of relative reality.

In its physiological aspects our tableau looks this way.

A sense datum depends on the organ that is being excited, not on the excitant. But the excitant, in this case the physical environment, is a constant, whereas the sense organ that is being excited, undergoes a refinement of structure and function. The reaction or sense-datum also changes and grows complex; its relation to the environment, too, undergoes complications.

Because of this, the signs or sense-data move from a full relative

---

[1] Helmholtz's view that signs in general correspond to the outer processes or objects is wrong. Such a simple or direct correlation belongs to only tactile signs.

objectivity to a full absolute subjectivity. In other words, they tend to a growing specialty of trait and diminishing verity. Their correlation with the state of physical milieu more and more turns away from the basic relative precision of primary signs.

But apart from signs of sense-data there is a multitude of others: signs-feelings, signs-ideas or their aggregates, which habit and heredity have transformed into a visibly simple and indissoluble symbol. These signs are indicators of still more complex changes of milieu, of sense data of a higher realm than sheer physico-mechanical happenings.

The sum total of all of the signs described above is the world of objects.

And if the signs corresponding to more and more minute and complex changes of milieu – our sense–data – show diminishing certainty and accruing specialization, this is precisely reflected by the situation in science. The latter's contents being objects or sense-data, its trend, is diminishing generality as issue of growing complexity.

Thus, in the very classification of outer events according to the degrees of their relative objectivity lies the defining of the measure of philosophical value in science. This value is in accord with the measure of actuality and generality of science's objects.

Really, if the simplest and the most general of the outer events – the mechanical – hold only a relative objectivity and have but a rather distant and indirect bearing on the understanding of the *Truly Existent*, then what absolute worth can be allotted to sciences studying events of still lower degrees of relative objectivity, still more complex and specific.

The conclusion arrived at by this long, fatiguing *exposé* of the data owned by the scientific theory of knowledge and of their general meaning, this conclusion is simple and highly important.

If sciences are at all capable of presenting a theory of knowledge, epistemology of a kind, their gift would still have no philosophical value. In the first place, such theory is of a relative nature by the token of its viewpoint. Secondly, if cosmos were such as nature science depicts it, even then the scientific facts supporting this theory of knowledge would have a more or less indirect, distant and totally unknown to us relation to the *Real Existent.*

V

In the above considerations it has been shown that of the outer events the mechanical are marked by a higher degree of relative reality.

But the highest belongs to the forms of the phenomenal world: space and time. These are known to even that realm of animated beings where life hardly sways. Without an awareness of space and time, no matter how feeble and dazed and dark such an awareness be, the perception of even faintest sense-data and struggle for existence of the most primitive kind is unthinkable.

Through these forms of the outer world, object and subject, milieu and perception, merge. Only visibly are these forms the beginning and the basis of perception. Actually they are the immanent and transcendental forms of Cosmos. Within their riverbeds flows the world's history.

The awareness of this is the philosophical view of the world; here lies the spring of the philosophy of knowledge.

And now, we shall turn to the questions: what can science give to the philosophical study of knowledge? Wherein lies the philosophical value of science?

Here we stumble upon a remarkable circumstance.

While scientific gnosticology feeds only on data from empiric and inductive science, only mathematics have never served as material for the former. Of course, some stray facts of mathematics could be used this way, being at the same time data of psychology. But in its specific nature, mathematics lend nothing to the *scientific* analysis of knowledge.

But in philosophical gnosticology the situation is reversed. Nature science gives her nothing *directly*: only rare data may have an oblique and specific meaning via metaphysics.

Pure mathematics, however, *as such*, constitutes the very basis, the very material for the philosophical theory of knowledge. We find this discovery first in Hume's reasoning, then in the famed querry of Kan't invoked by Hume's thought: how are synthetic *a priori* judgements possible?

This difference of bases for the two doctrines of knowledge makes clear their relationship. Scientific or empire gnosticology describes the origin and development of empirical knowledge as conditioned by the actuality of the outer world and fed by nature sciences.

Philosophical gnosticology or epistemology describes and defines metaphysical knowledge, that conditioned by the transcendental and ideal quality of Being.

In defining such knowledge, philosophical gnosticology exploits pure mathematics which grants us at least the insight into the *a priori* determinants of the outer world's forms. This is why, scanning the data which science may give the philosophical or speculative doctrine of knowledge, it would prove really sufficient to look into what precisely mathematics bring to this theory.

Pure mathematics is usually divided into arithmology, study of discontinuous functions, analysis, science of continuous functions; and geometry. We need, however, a division guided by a different viewpoint, that of philosophy. With this in mind, pure mathematics should be described as algebra (in a broadened sense) with its subdivision – arithmetic. The latter's objects are aggregates of visibly separate spatial objects; distinct points intimated by a visioning of aggregates of moments of pure time. Geometry's objects derive from contemplation of space.

Both in aim and general method, algebra and arithmetic are similar sciences. In a mathematical situation they study the possible quantative permutations deductible from the given premisses.

But in arithmetic the quantative terms of a problem are concrete and uniform: numerical amounts made up of identical units. Their permutations result in a simple, concrete and unconditional figure. In algebra, the basic premises are abstract and multiform: elements not reducible one to another x, y, z . . . a, b, c. Therefore algebra presents only such transformation of premisses through which they would come into simple relationship. Even so, the result is abstract, complex and conditional.

The immediate aim of that part of mathematics which we called "algebra"[1] is the simplest solving of a given mathematical problem; that is, such marshalling of mathematical premisses made up of algebraic terms and equations, that would set them into a simplest relation, and therefore, lead to a solution.

Thus, for example, in differential equations one separates the variables to make the integration easier. The variety of ways in algebra depends only on the fact that problems of a peculiar nature demand, for the formulation of their conditions and for a precise solution, a variety of mathematical concepts: function, derivative, differential, probability, variation, etc.

The aim is the same, as in elementary algebra: to bring the premisses of the problem into the simplest relation.

The words of Carnot about differentials throw true light on the heart of the algebraic operation.

"Differentials enter into calculation only as auxiliary figures; and this calculation is considered completed not before those auxiliary symbols are eliminated".[2]

Because number, too, may enter into a given problem as its mathematical premiss, the theory of numbers appears in this sense to belong to algebra.

Even the simplest actions of algebra (addition, multiplication, extraction of root) are transmutations binding a group of quantities into one. But there also is a variety of special, complex transmutations: Newton's binomial, for example, or the expansion of trigonometric and logarithmic functions into series based on operations pertaining to indeterminate coefficients.

In higher mathematics, taking the first examples that come to our mind, one could point to the differentiation of logarithmic functions

---

[1] In our sense pure algebra comprises: the so called elementary and the higher algebra; differential and integral calculus minus their application to geometry; calculus of variations and of finite differences; the theory of probabilities and, to some extent, even the theory of numbers.

[2] "*Elles n'entrent dans le calcul que comme auxiliaires et ce calcul n'est regardé comme fini, que du moment ou ces auxiliaires en sont toutes éliminées.*" Carnot: *Réflexions sur la méthaphysique du calcul infinitésimal*". *Chapitre* I, 1.

($y = \log x$), based on the evaluation of $(1 + 1/n)n = e$ (basis of Napier's logarithms) when $n = \infty$; to the integration of parts and by substitution; or to the introduction of the integrating multiplier.

Theorems themselves are formed of transformations linked by ties of logic, but ones of a more general order. With given premisses and a private proposition to be drawn from them, mathematical dress and procedure added to such an action make this a transmutation. If from the given mathematical conditions or premisses a general thesis is to be evolved and strong logical seams sustain the mathematical syllogism, then we have a theorem of algebra.

In the old currents of mathematics, algebra was but an assistant in the solving of problems of geometry, mechanics and other concrete quantative tasks.

In the new mathematical schools, algebra tends to become a purely combinatory doctrine: a science of permutations.

This is why the work of mathematicians, such as Gaulois, Sophus Lie, etc. was directed toward a creating of methods of aim-aware transmutations, toward replacing the art of improvisational permutation by rigid scientific procedure.

This tendency of mathematics to study all possible transformations in any groups of quantities corresponds to the urge of nature sciences to scan everything liable to become their subject: from spots on celestial bodies to an infusoria of which there are millions in a drop of water.

VI

And so, the aim of algebra is not the solving of problems of quantities in an algebraic way. Out of such problems algebra grew, of course. But its fundamental, purely theoretical contents lie in the study of possible changes or alternatives in the quantitative formulation of conditions or premisses.

Algebra's method or the nature of her reasoning is transmutation of formulae and equations.

Turning to the philosophical worth of algebra, let us note that as material for problems of epistemology, it is of no lesser import than geometry.

Algebra could reveal philosophical truths of great weight. But as fas as we know, it has never served *this* realm of meditation. As to mathematicians, since Leibnitz they ceased to be philosophers in a true sense.[1]

On the other side, arithmetic still enjoys some attention from the philosophers, although beginning with Kant, they all chewed over and over the thesis of arithmetic's essence being continuity and visioning of pure time. Nothing was ever added to the old motive.[2]

---

[1] Minding only the scientific value of mathematical methods, some of the mathematicians sport but slender philosophical equipment. Such is, for example, Carnot in *Réflections sur la Métaphysique du Calcul Infinitésimal*.

[2] Schopenhauer, at least, feeling that he cannot say much about it, grants only a dozen lines to the subject. See *Ueber die vierfache Wurzel des Satzes vom Zureichenden Grunde*.

No one ever noticed that arithmetic's blood kinship with algebra lies not in its being an organic study of quantities, but in its vision and analysis of all feasible transmutations within formulated groups of quotients.

The higher philosophical value of arithmetic's peculiar aspect as a case of purest mathematics, has been lost sight of. Helmholtz, for example, pays attention only to the psychological roots of arithmetic; is occupied by the questions of the origin of numbers, arithmetical axioms and so forth; never bothers about the numbers' deeper spring. This situation is even more pointed as to algebra: its aspect *sui generis*, procedure, nature of reasoning, do not as a rule engage the mind of philosophers.

Space allows positions to be defined by concerted groups of geometrical units. In a similar wise, in the moments and groupings suggested by our vision of time lies the way of defining numerical quotients by sets of algebraic elements.

To elucidate and specify this thesis and to draw its philosophical implications, I beg to state, that in addition to being a study and accounting of all feasible transmutations in formulated relationships, algebra, the most generalized form of counting, acquires the same characteristic traits that numbers borrow from time: continuity, one-dimentional nature, contemplative abstraction.

As soon as we proove this, the proposition that *in its roots* algebra is contemplation of pure time, can be sustained.

In algebraic chains of reasoning and permutations, their continuity admits of no leaps. Every stage of transformation is conditioned by the previous one and, in its turn, conditions the next. This continuity is not of a logical or syllogistic order. The logical element does enter in the variable part of algebraic reasoning: in the setting of a problem and the writing of its equation.

The one-dimentional nature of arithmetic is irrefutable. But just because the philosophers who have never analysed algebra, know one-dimentional traits from the simple reasonings of arithmetic, they have convinced themselves that the trait of one dimension is incompatible with multiple relations. This is the opinion of Schopenhauer, for instance.[1]

But let us see where the variety of relations in algebra comes from; which aspect of it these relations point to; and where they lead us.

In all branches of algebra – and in the calculus of direct functions as well as of the indirect – not only the variety of elements but also, the multiplicity of relations, that is, concepts the nature of which algebra studies analytically, flows from the multiplicity of conditions (or premisses) of the problem.

As elements a, b, c, ... pointing to definite quotients, or elements x, y, z pointing to not yet defined ones, the concepts of the differential,

---

[1] Schopenhauer: *ibidem.*

integral and terminal difference, of probability or variation pointing to particular quantities or peculiar relationships of quantities, – all of these were born from the necessity of expressing conditions or premisses of a problem.

It follows that multiplicity of relations is not the essential or typical trait of algebra. The latter is mainly characterized by its generality and particular nature of its reasoning. This is why, in spite of all the simplicity of the elements of arithmetic, and in spite of the uniformity of the latter's relationships, the character of arithmetic's reasoning makes it a species of algebra. This is also the reason why all the variety of algebra's inner relationships does not in itself cast doubt on its one-dimensional nature.

Finally, what, in the last count, is an algebraic proposition, and how does it signify a given stage of transmutations?

First of all, it means a concrete quotient either defined at once or set in relation to a certain series of quantities. An algebraic proposition presents every quantity as a function of a group of constants or variables, and, in the last count, it means a definite quotient: in a private case – either a definite number, or one to be found in a definite series.

This proposition extends also to specific quantities, such as complexes of the form $a + bi$. These symbolic quota are resolved into concrete ones not directly but through their link with real quantities. Thus Euler's formula gives their connection with trigonometric functions and the basis of Napier's logarhythms e. Their modules or variants are certain vectors of a definite length and direction.

If, therefore, every algebraic expression is, in the final count, reduced to a definite quotum; if every stage of transmutation crystallized in that expression, presents, in essence, a concrete quantity – then algebra is, in spite of all the variety of its inner relationships, one-dimensional.

VII

The third trait of algebra is the speculative one – or shall we call it "contemplative" – interconnected with its being synthetic *a priori*.[1]

As to this trait, the philosophising mathematicians confuse the problem of the anthropological roots of mathematics with that of its emerging from a speculative spring. Helmholtz, by the way, who is considered by some a sort of Kantian, did not fully understand Kant. Transcendental aesthetic is not an antropological doctrine.

For us, it is not at all important to prove that mathematics – in a private case, algebra – has been born from *a priori* speculation.

We do not even maintain that mathematics emerged prior to ex-

---

[1] At times, speaking of these traits, we have in mind not algebra alone, but mathematics in general. The reader will surely find no trouble in extending the above considerations to geometry when we come to it, and spare us repetition.

perience: only, that the analysis or, better, the exausting of possibilities in space or time relations – which is all, mathematics appear to be – *could be* achieved prior to or passing by experience. For mathematics, the purely speculative experience that is potentially given in the outer world's transcendental forms, is sufficient.

Therefore, we shall try to show that mathematics *could be* created by sheer speculation *a priori*, even if its actual origin is empirical and through induction.

Our second task is clearly to tell the contemplative nature of mathematical concepts from the speculative traits of mathematical reasoning. As an issue of the former, mathematics is *a priori*; in connection with the latter, synthetic.

Our first task is sustained by the conviction that the forms of mathematical objects exist prior to experience and shape the experience. And this, no matter how we explain it: scientifically, that is, empirically and evolutionally; or philosophically – in idealistic terms.

As to myself, I adhere to the conviction that the transcendental and ideal forms of cosmos preceed and shape it. For an admission that space and time emerge, like any other sense-data, as a result of some organ's function, or a subtler concession that sense-data are formed through subconscious recording of muscular motions accumulated through heredity, these beliefs could only lead us to absurdities sourced in the already stated vicious circle.

The above beliefs are but a *scientific* truth, conditional and relative. It can have no weight in philosophical, that is, absolute epistemology.

Applying the scientific "truth" about the origin and meaning of space to our problem, one falls into Helmholtz's search for psycho-physiological roots of mathematics. Even if he did prove that mathematics have an empiric origin, this would have not the slightest bearing on our *philosophical* problem of space, nature of mathematical reasoning and so on.

But again, it is true, that the experience of our senses could be – and at times has been – the inciter of mathematical concepts and perhaps, also of mathematical reasoning. It is easy to see, however, that even if its basic concepts were formed empirically, mathematics has long since abandoned that deceptive, filmy bridge: the experience of senses.

It is highly significant that an equation embodying a law of nature is chosen from the ocean of mathematical propositions, not *vice versa*.

Using the procedure by which Helmholtz has demonstrated the feasibility of non-euclidean planimetry, one comes to a round about way of sustaining the proposition that mathematical concepts are not necessarily tied to our actual sense-experience; that mathematics *could be born* of *a priori* speculation. Helmholtz suggested thinking beings with two-dimensional bodies living on a rigid curved surface. Their geometry should be non-euclidean. What I propose, is no less feasible or imaginable.

Let us fancy a reasoning being with most varied and limitless mathematical abilities and a faultless mathematical judgement. If in the imagination of such a being, by some cause, all his mathematical concepts were aroused so that they presented themselves lucidly to his awareness, all mathematics would have been born at once – through all feasible permutations in the swiftly emerging, novel elements of mathematics.

This means that the march of mathematical thought *should* tend to the ideal in the same degree as actual mathematical creation tends to perfect formulation. Actually, the growth of mathematical creation is limited by the imperfection and narrowness of mathematical thought *only*. For the necessary experience is always there, in the brain of mathematicians, waiting to be created beforehand.

In the history of mathematical thought it happens precisely as just described.

Let us take, for example, the history of the all embracing equation of mechanics, the generalized Lagrange principle of feasible displacements.[1]

The principle was known long before – to Galileo who evolved the terms of the lever's balance from it. And it had taken two hundred years of development of analysis and calculus of variations – to enable Lagrange to formulate and generalize his principle with all precision. Had Galileo created analysis himself and brought it to perfection, he surely would have established at once the Lagrange principle in its strict and general form.

Thus the *a priori* unfoldment of mathematics, as one can see, lies in the fact that, unlike in other sciences, the data or experience needed by mathematics is given or created beforehand. Its development does not have to be fed by sense-data. The latter, and even the scientific urge, are only inciters of mathematical work; are but conditions necessary in view of the inertia and limited potency of actual mathematical thought yet basically inessential for mathematics as such.

There remains now to speak of the speculative traits of mathematical concepts, traits linked to their *a priori* nature as against the speculative traits of mathematical reasoning linked to mathematic's synthetic mark.

Algebraic concepts present certain pure relationships. If algebra were nothing but a sheer drawing of contents from mathematically formulated premisses, even this set of limited acts of deduction would be of a purely speculative nature; could exist prior to outer experience.

But the *contents* of so abstact a relationship are yet evolved through

---

[1] A system will be in balance if the sums of energy of forces applied through the length of the systems's possible displacements will equal zero. Applying this principle to forces lost in overcoming resistence, Langrange evolved his famed equation for any case of motion and any system of points: $P\delta p + P'\delta p_1 + P_2\delta p_2 + \ldots. = 0$; or $P\delta p = 0$, where P, P_1 and P_2 stand for lost energy, and δp, δp_1 and δp_2 – are elements of possible displacements.

reasoning operating with traits visual and contemplative. Only figuratively speaking can one insist that algebra dissolves its concepts analytically.

Foretelling a probability, for instance, is an unfoldment of some pure relationship. It is but an evolving of a mathematical expectancy which is a relation of a number of some elements in a certain series to the entire number of its elements.

Aside from its immediate qualities, the speculative nature of such algebraic relationship dwells in the absence of a correlative in one's experience. Yet the relationship itself is neither abstract, nor discoursive; it is visual. But again, the reality of sense-data is meager in both – mathematical sense and comparison with the utter variety and utter refinement of algebraic concepts.

In more complex algebraic relationships their purely *a priori* nature is unmistakable: in such concepts, for instance, as derivative, integral, variation.

As to the speculative order of algebraic judgements, which are but stages or links in the chain of algebraic transmutations, one may note that if their synthetic nature is still faintly seen in simple transmutations (addition, multiplication, division), it grows more visible in complex algebraic acts such as extraction of root, raising power, or logarhythmic operation. In still more complex transmutations, their synthetic nature can be hardly disputed.

In algebraic reasoning and in that of mathematics in general the apodictic truth and generality can be affirmed *ante rem* – on the strength of axioms; or *post rem*, by a possible checking with experience.

But the axioms of algebra are but judgements concerning algebraic elements or concepts, and judgements without premisses.

If the objects of mathematics are not concrete, which is so manifest in algebra; and if the reasoning of mathematics are not *a posteriori*, then, by this token alone, the truth of mathematical judgements can but does not have to be sustained by experience.

The measure of *this* truth is purely and expressly mathematical: a strict continuity of reasoning. No matter how much experimental checking up or faultless experimental application confirms a mathematical proposition, it will not be a mathematical truth (though, in this case, it may be an experimental truth), until it is sustained by a precise and consistent set of mathematical premisses.

No matter how clear it is that of all the isoperimetres the arc of a circle encloses the maximum and minimum surface; or that the equation $x^2 + y^2 + z^2 = 2x + 2y + 2z - 3$ is correct when x, y and z are equal to 1, one can not consider these propositions mathematically true, until the variation of a definite integral proves the first, and a clever transmutation – the second. The truth of these propositions can be confirmed in no other way than by showing that the judgements are correct *mathematically*.

It is exactly the same in geometry. If, for example, we would measure the relation of the circle to its diameter with "ideal" instruments, using an unfolded and most precise circle made of ideally thin wire (which was roughly done by Archimedes), we still would not obtain a *geometrical* truth. Only the *mathematically* reasoned calculation of Π can give us such truth.

The verity of a mathematical thesis can be tested only by its *special* criterion. This is why such truth is apodictic.

Here is all that one can say of the philosophical aspect of algebra.

VIII

The other part of pure mathematics, that dealing with positions in space, is geometry.

As to euclidean geometry, it is important to mention only that there exist geometrical proceedures neater and more proper than those of Euclid. His proving ways are rather logical – through abstraction; not geometrical – through visual procedure.

There are geometrical methods of much purer rule: in the so called "problems of construction" or the method of geometrical *locations* created by the school of Plato. These speak of the speculative and synthetic nature of geometrical judgement better than Euclid's reasonings.

But at any rate, as to the philosophical aspect of euclidean geometry, it is hardly possible to add anything essential to that said on the subject by Kant and others after him.

Then again, in problems of epistemology it is the non-euclidean geometry which is now of supreme importance.

A geometry built on the discarding of the famed eleventh axiom (the *postulate* of Euclid) denying its absolute truth, which Lobatchĕvsky did, – such a geometry is free from inner contradiction. And is valid for a surface of a certain negative curvature, the so called pseudosphere of Beltrami.[1] Which means that Euclid's geometry is but a private case of geometry for surfaces admitting a uniform formation.

Infinitesimal parts of a sphere are flat surfaces, and Euclidean planimetry is correct in their case. In this sense one may call such planimetry – and with more justice than application of analysis – differential geometry.

In spheric or pseudospheric geometry many propositions of Euclid are, indeed, out of place: of his axioms – the renowned postulate; of his theorems – the one concerning the sum of angles in a triangle.

For in a spheric triangle this sum is larger than two right angles; in pseudospheric – smaller.

It again follows, that Euclid's planimetry is not of a general validity; it is merely a private case of geometry of surface.

Extending these considerations to solid geometry, one arrives at

---

[1] The curvature of surfaces is characterized by a certain constant the symbols of which signify: positive and negative ones – a spheric surface; mixed curvature – a pseudospheric. Zero curvature denotes a flat surface.

two suppositions. Firstly, space where, in Riemann's terminology, inner measure – relations others than in the space covered by Euclid's solid geometry, are possible. Secondly, the number of dimensions marking euclidean space is only a private case; this number can be larger.

Which means that space can be spheric with positive or negative curvature; in such space the shortest distance between two points would be a curved, not a "straight" line. Also, if euclidean space is triply expanded into a multi-dimensional continuum – as Rieman formulates it, then we have here a private case of the n-dimensional space continuum.

Much coarser in a way are the formulae of non-euclidean space offered by newer mathematicians: Poincarré, Sophus Lie, and others.

Defining space as a potential of transplacements, one perceives that of all feasible groups of changes in space position for a rigid body, only a part of them can be effected within the euclidean space. Therefore some groups of displacements are possible only through a surmising of a space of another kind.

Calculating the sum of angles in some immense cosmic triangle, the parallax of stationary stars, for example, mathematicians tried to decide whether our space is flat or curved.

Lobatchĕvsky found such parallax negative, and this could mean that our space is pseudospheric. But the fact is more naturally explained by the faultiness of astronomic instruments of those days.

In fact, these direct measurements used for the defining of our space's curvature, have no significance at all. No matter how tremendous the distances taken for measurement are, they would be infinitesimal in relation to the immense expanse of the universe. And no matter what the real curvature of our space is, for the limits of man's measurement – differential geometry will still be valid.

Philosophical conclusions drawn by mathematicians from the new geometrical concepts contain much confusion. In geometrical matters particularly, philosophising mathematicians mistake the generalization of a mathematical problem for the deepening of a philosophical. And like in algebra, they confuse the question of the actual origin of geometry with that of its emergence by way of speculation *a priori*. That is, they confuse a problem of philosophy with that of anthropology.

And finally, as to the question of whether a geometry other than Euclidean is "right" or not, the problem posed by Kant demands no solution. The latter pertains only to the question of how the *euclidean* geometry has come about. This can not be answered in a roundabout way, by showing that other geometries are feasible, and that we know the present one only because it is related to our outer experience.

Suppose we leave at rest a number of "philosophical" misunderstandings and even absurdities springing from the proposition about the *concrete* possibility of n-dimentional or curved space. Let us then briefly analize only the inner meaning of the new geometrical concepts.

1. To prove that *our* idea of space as a three-dimensional continuum is concrete and *a posteriori*, mathematicians following Riemann, pointed to still another, and in their opinion, a brighter example of such a continuum: to the system of color as a dependency of three variables. These are strength of light, color, density. But the continuity and the three-dimentional nature of this continuum are doubtful. Namely, because outwardly every color shade corresponds to a certain concrete and rational number of vibrations, and is therefore a non-continuous function.

As for the three-dimensional nature of the color system, it does not really resemble the articulate three-dimensional frame found in space. Color's dimensions merely drown one in another.

2. The statement that non-euclidean geometry shows a possibility. of frames of space other than those known to our perception, is not valid as to the geometry of surface. For, in the first place, the study of curved surfaces can use, even if only in a lame way, the methods and concepts of analytical and differential geometry (application of analysis) whose formulas presuppose euclidean space. Secondly, euclidean planimetry is valid not only for flat surfaces but also – for the flat euclidean space. Therefore, a repudiating of the generality of planimetrical axioms by showing the feasibility of another geometry does not sustain a denial of these axioms' universality for space: not until the feasibility of *stereometric* frames other than ours can be demonstrated.

3. It is possible to think of space with a curvature other than zero as a medium of a certain peculiarly staggered density. Linking our geometrical concepts to the mechanical – which one of the fathers of new geometry, Lobatchěvsky, wanted to do – one should deem straight any direction taken by a point or an infinitesimal of matter moved by a suddenly acting force, no matter how the latter is directed and in a milieu of uniform and constant density.

It is clear then that in a milieu of density continually increasing or diminishing by infinitesimal degrees, and distributed in concentric layers of infinitesimal thickness, a point or molecule "pushed" by any sudden force will take as its direction a path of positive or negative curvature.

Something similar takes place in astronomic refraction. If the density of the atmosphere were decreasing by infinitesimal degrees in infinitely slender layers, the refraction would be arc like.

Linking geometrical concepts to mechanical, we thus perceive that geometry of curved spaces is but kinematics for environments with peculiarly organized densities.

4. Space with a non-zero curvature must be either "bordered,"

that is, in some way circumscribed, as in Lobatchĕvsky's image, or finite, as in Riemann's. But we can *actually* imagine space only as "borderless" and infinite: a vision of space that could be in no way based on experience! Of necessity, we merge the concepts of the "borderless" and the infinite into something that could be born only from the *a priori* idea of euclidean space.

The above considerations make it clear that objects of non-euclidean geometry alien to the frame of "our" space with their inner measures and relations, are imaginable only through a translation into terms of euclidean space, makeshift as such translation may be.

Those non-euclidean symbols must be revamped into mechanical peculiarities of a definite environment. Then, for their interpretation, they need no concept of another medium, intangible and unimaginable. As for the frames of non-euclidean space, it must be circumscribed or conditioned which again contradicts to our native concepts.

5. It is impossible to *actually imagine* space forms of more than three dimensions. Figuring out how will an object of more than three dimensions appear to our inner vision, we really but reshape the curve and expanse of the dimensions in common use. A space form of more than three dimensions can not become an object visible to us or touchable. Which means, it is neither imaginable, nor realizable.

One could argue that molecules of physical matter or the immense stellar spaces are not perceptible either.

They are, however, *imaginable.*

Still, we can not fancy a change in our mode of vision which, *without contradicting its nature,* could give us the ability to perceive or to imagine space forms of more than three dimensions. That is, we can not imagine a concrete possibility of such forms, or *concrete* conditions for their perceptibility.

6. If bodies of more than three dimensions can not be objects of our direct perception and knowledge, then non-euclidean geometry has no meaning for epistemology, particularly, for the problem of the universality of the statements of euclidean geometry. For epistemology probes the latter's universality as related to empirical existencé, that is, to the world of objects or of what can become an object.

If it is so, new geometry does not affect the question of where does the above universality come from; this querry pertains but to objects of *our* kind of perception.

7. Propositions of metageometry are valid and universal for surfaces while euclidean planimetry is but a private case of a feasible non-euclidean geometry of surface. But this in itself does not prove that the axioms of Euclide's planimetry are not universal: its statements are entirely valid for Euclide's solid geometry, and therefore, for *our* space and *our* empirical world.

8. How is non-euclidean geometry "possible?" In which sense?

As to geometry of surface, this question can be answered, *ceteris paribus*, in the Kantian way: non-euclidean geometry of surface is no new fact that compells a revision of concepts. As material for our present philosophical problem, non-euclidean geometry of space with other inner measure relations than those in our common use, should be viewed as mechanics for certain peculiar environments, rather than mathematics.

Even if geometry of curved space should be treated as pure mathematics, it has to be considered but an algebraic doctrine of transmutations of euclidean geometry's theorems where a certain vector $r\sqrt{-1}$ is introduced.

9. Finally, we have with us geometry of many dimensions or the theory of multiple continua.

Again the latter is in essence an algebraic science of feasible transmutations in certain systems of equations with many variables, which, for attractive analogies, one is tempted to interpret as geometrical structures based on many co-ordinates in multi-dimensional space. The meaning of such geometry is purely algebraic: as material for and possibility of certain general solutions for the theory of continuous groups of transmutations, and for that of differential equations as well.

Thus the question of how is geometry of many dimensions and non-euclidean geometry in general possible – in as much as these doctrines belong to pure mathematics – this question really amounts to: how is algebra possible?

The philosophical analysis of algebra compels a conviction that Kant's solution was right. That of non-euclidean geometry, too, sustains Kant mightily.

IX

Substituting algebraic elements and correlations for geometrical or for nature elements, and unfolding the algebraic contents of thus formed ideas or equations, we obtain the application of algebra to geometry, to pure mathematical and pure nature science. Every transmutation of this kind presents a new relation between the elements of a proposition. When one grasps the meaning of this new relation, geometrical or physical, we have a new theorem of geometry or a new law of nature.

Analysing the equation for the ellipsis, by an algebraic reshuffling of its formula, for example, one finds that the geometric location of the middle in parallel chordas is the diameter of the ellipsis, and that circle is its limit. The very equation of the ellipsis is arrived at by a tortuous transforming of the equation showing that for every ellipsis, the sum of its two radii-vectors drawn from the same point is a constant, namely $c_1 + c_2 = s$.

In abstract nature science the transmuting of its equations makes it possible to obtain mathematically laws known from experience; and particularly so, when one eliminates the mechanical moments bound to geometrical ones: in the theorems, for example, on the center of weight or on the motion of the center of inertia.

Thus, in analytical mechanics, transmutation of the basic Lagrange formula brings us the law of squares. Applied to celestial bodies, it results in the second principle of Kepler obtained empirically.

In the kinetic theory of gas, out of the key formula for gas pressure obtained by sheer analysis, the laws of Avogadro, Dalton and Mariotte-Gay Lussac can be deduced. These laws were first arrived at empirically, and then sustained mathematically.

As for their philosophical aspect, in trying to find to what extent abstract sciences can serve as material for epistemology, one has to mind the following points.

Setting in algebraic symbols the elements of a geometrical figure or body, we can – through algebraic reasoning – find some of the figure's hidden geometrical traits. In its very calculation mathematical action seems to tend to pure reasoning of one kind, algebraic or geometrical.

Of this, the Menehm problem – doubling of a cube – is a telling example.

Here we see a geometrical problem that is treated to geometrical reasoning and is brought to a geometrical solution. Yet the *analytical* path leading to the latter remains purely algebraic.

This occurs in an even sharper manner in that tremendous march of transmutations that evolves the equations for the ellipsis and the parabola. From this alone one can perceive that only in an agreed way, only conditionally and temporarily – for the sake of argument, as it were – can algebraic symbols be substituted for geometrical elements. Essentially, the chain of reasoning tends to remain either strictly algebraic or purely geometric.

Saying it in a traditional Kantian sense, an argument with such a dual underpinning is really an issue of one source at a time: the visioning of either pure spatial positions, or of pure relationship of moments.

One of the methods of application of algebra uses elementary functions and builds its formulas on a system of co-ordinates: the way of analytical geometry. To insist that such operations are algebraic, is clearly a misunderstanding. Yet there are mathematicians doing so.

True, the algebraic method used by analytical geometry is neat and fruitful. It is quite possible that one could not unveil certain hidden traits of some of the curves by sheer geometrical operations. Still, algebra plays here a definitely secondary role. It is idle to insist on analytical geometry's having an algebraic meaning of its own, when the various symbols of its formulas point to spatial direction: the variables – to parts of some straight lines; the coefficients – to tangents of some angles; the given term – to the initial co-ordinate.

These observations pertain also to the geometrical study called the calculus of positions. Mathematicians sometimes call the later geometry minus space.

Speaking of analytical geometry's propositions, one must note that although certain details of the nature of curves were known before their analytical discovery – Appolonius, for example, knew of some traits of conic sections – this does not mean that analytical geometry rediscovers *a priori* all that it has previously learned *a posteriori*.

The *a posteriori* knowledge is not necessary, since analysis can evolve *all* the traits of a curve from its equation.

As for the synthetic nature of analytical geometry's propositions, it is linked to either of that geometry's aspects, algebraic or geometric. The speculative essence of these judgements reveals itself also in the fact that analytical geometry fruitfully exploits the idea of geometric location: setting up, for instance, the equations for curves.

Pure nature science is the sum of those synthetic *a priori* propositions which are general and valid as to the contents of the outer experience or sense-data, while pure mathematics pertains to the *forms* of experience alone. Kant saw the basis for pure mathematics' validity for the world of sense-data in the transcendental categories of reason, which exist prior to experience and organize it.[1]

Refuting these categories as principles unstable and even doubtful as to their autonomy, we are left with only one kind of "pure nature science": only one the universality of which and apodictic truth are not merely logical or issue of doubtful "categories," but mathematical, originating in actual forms of knowledge. Which means, that only applications of mathematics have the right to call themselves "pure nature science."

Not all applications of mathematics, however, deserve such name. Even in pure mathematics only kinetics present a clear and strict example of pure nature science.

Portraying pure forms of motion, itself a function of pure elements of space and time, kinetics is not really pure mathematics. But by the same token, kinetics is the study of the pure contents of matter in as much as matter is the "moveable in space," and an issue of space and time. Thus kinetics contains not a single empirical element of nature and unfolds its contents using the four co-ordinates of space and time. Where, however, the empirical element can be linked with the mathematical in a correct way, kinetics approach nature science.

Varignon's graceful theorem about the momentum of the sum of forces builds the momentum of each force as the surface of a triangle, where the base is a straight line equating the force in quantity and

[1] Kant: *Prolegomena zu jeder Metaphysik* (§§ 14–35).

direction, and the height – the perpendicular to this straight line or of its continuation, dropped from a point in relation to which the momenta of forces are set.

In celestial mechanics, presupposing that all the mass of a planet is "condensed" at its center, one affirms the dependence of the trajectories of celestial bodies solely from the relative position of their centers of matter and amounts of mass: that is, from certain moments reduced to purely mathematical terms. In analytical mechanics the same supposition permits the deduction of the principle of motion for the center of inertia.

In other application of mathematics and mechanics (to nature sciences, psychology, statistics, logic) the basic elements are drawn from even wider empirical data, physical or mental, but their contents are being evolved through mathematical ways and reasoning. The judgements so obtained must present apodictic truths, as mathematical propositions do. But they are universal only in as much as their terms are being empiric ones.

Juxtaposing all these remarks on the application of mathematics, we see, that its *purely* mathematical varieties as well as the application to nature sciences, have the same meaning and value for the philosophical theory of knowledge as pure mathematics do. In such strict application one perceives even better how sternly speculative its reasoning is, algebraic or geometrical. In application to nature doctrines, the judgements or propositions are *a priori* and apodictic *as far as* they are mathematical.

But what of the philosophical value of mathematics itself? Where exactly lies its significance for the discipline, it is called to serve as material: the philosophical theory of knowledge, or epistemology?

A review of the foundations of mathematics convinces us that not only Kant's solution but his reasoning, too, is entirely right. More than that, such conviction is sustained by either, the probing of the roots and principles of mathematics, or of the direction of its progress.

Even being moderate in attributing high importance to the new successes of mathematical thought, one has to agree that new mathematics have assumed the correct aim. That is, algebra straining to become the pure science of transformations, a forecast of all feasible trans- and permutations in any algebraic formula of a problem; geometry tending to be *generalized* science of space which it has attained at least in the non-euclidean geometry of surface.

The review of the philosophical value of algebra has shown that the transmutation moment is the very one to be linked in some way to pure time. In all evidence algebra borrows the traits and instrumentality of the latter: its one-dimensional nature, continuity, visionary quality. Thus it follows also that arithmetic is just a form of algebra.

Epistemology should base itself therefore on analysis of the latter.

A sounding of the foundations of geometry shows that new geometry does not really set before us *new facts* compelling us to alter or at least revise our philosophical convictions. For geometry of many dimensions is a kind of algebra with a special meaning for the science of trans- or permutations. As to geometry of curved spaces, it is in one way an algebra (with its pith – a study of transmutations of Euclidean propositions altered by the introduction of a peculiar vector); and in the other way, mechanics for some special physical environments.

Non-euclidean geometry of surface is not that *new fact* either. It does not really demand new geometrical concepts; it can be reached through paths speculative and *a priori*. Which means that the philosophical significance of non-euclidean geometry is the same as that of the euclidean. Together with algebra, both branches of geometry re-affirm the hardly shaken conviction that mathematical knowledge arrives or may arrive prior to the experience of senses; may be evolved from the peculiar, private experience given by the vision of pure forms of cosmos. Mathematical knowledge is an exhausting of the contents of these forms: bare relationships in time, pure positions in space.

In this way only is the speculative nature, the universality, the apodictic truth of mathematical judgements possible.

Relating this thesis to its sources we come to a firm conviction that pure transcendental forms of cosmos must exist prior to it, as conditions for its possibility: potentially, in the transcendental subject, the carrier of the Universe. Upon this is the transcendental ideality of Being conditioned; upon this is based the co-presence of the former trait and the universe's empiric reality.

Such a high deduction from epistemology is arrived at at even an early probing by reason of its own limits and of its own abstract use. From this proposition the synthetic *a priori* nature of mathematics is deduced. It follows that the philosophical significance of mathematics, or of those of its branches which have a bearing on gnosticology, lies in that it merely sustains a purely speculative doctrine. Mathematics could, however, give philosophical analysis of knowledge details that might escape in an abstract critique of reason.

Thus mathematics has a real and absolute philosophical value, even if a limited and indirect one.

X

Our task – to define the value of science for two formal disciplines, the relative theory of knowledge and the absolute, is at an end.

This has not been a task purely methodological, that is, a search into what sciences in their possible contents can give the above theories. Rather it has been a review and critique of the *actual* contents of the sciences as material for epistemology. The review has shown that the possible scientific theory of knowledge – say, the Darwinist school, would not survive scrutiny and can not provide the basis for a philosophical view of the world. Defining and explaining the birth of sensations through natural selection, scientific epistemology of such nature already presupposes a universe, that is, a sum of sensations. Formally, by its higher viewpoint, the theory of knowledge must of course, be absolute; view the universe as an object; be essentially metaphysical, that is, consider the objective world of events in relation to the immanent nature of things.

However, even strictly scientific views do allow for a formal epistemology. Taking it, for example, that the frames of knowledge were formed gradually but hardened at a certain stage of development, one could harmonize transcendental aesthetic with Darwin's theory.

What remains to be discussed now, is the value of scientific concepts not to formal gnosticology, but to the philosophical view of the Universe. But even here, we confine ourselves to an establishing of how far can science-concepts serve as material for a view of the Whole; and how much can science-concepts advance such a view *through their own* power.

The value of a philosophical view of Being depends on the worth of its ingredients: ideas and concepts. It is the logical critique that opens to us the contents of those ideas.

And logical critique of the scientific concepts that form a scientific viewpoint reduces itself to the following. First, the outer or formal critique defining if a scientific viewpoint is generally possible; if an absolutely realistic one can be obtained; that is, whether one can hope to relate and harmonize in a unified scientific philosophy concepts from the various sciences; and whether one can do so by attributing to them absolute significance.

The other aspect of the above logical critique is the inner, essential one, the sounding out of the very pith of scientific ideas.

The formal critique of scientific concepts reveals the following.

α) A completely unified scientific view of the Universe is not really possible. Up to now it has proved unattainable to unite the two realms of Being, things of the matter and things of the spirit – in one general mechanical theory.

β) It seems impossible to arrive at an absolutely realistic, scientific philosophy, for it would be made up of a number of hidden antinomies.

An essential scrutiny reveals, for example, the absurdity of the dynamic concepts which should be the basis for a scientific keypoint unifying all sense data. Besides, these concepts have no purely experimental physical lining. It follows that a purely scientific yet thoroughly unified philosophy is impossible.

The impossibility of an absolutely realistic view of the entire universe treated as unit, can be easily established. Scientific concepts refute each other, and with it, that all important and unavoidable global view.

If the world of empiric existence were *absolutely* real, it would have to, as a potential of sense data, spark a definite sum of sensations at every moment: this, on the supposition that the world as a whole is confronted by a corresponding subject receiving sensations all at once from all things of the empirical world.

The antinomy of space lies, for example, in the concepts of the trajectory of comets.

Their path is a parabola. But even if it only seems parabolic, as some scholars think, being in effect but an elongated ellipsis, through the same laws of nature, of gravitation, for example, and in conjunction with Kepler's principles, which give the mentioned path an elliptic curve, it might as well be parabolic or hyperbolic, that is, one with two ends meeting in infinity.

Here lies also the antinomy of time. The latter is hidden also in other scientific concepts, such as calculation of periods in nature's physical and animal life.

For the phylogenetic history of animal beings it is not only impossible to set a definite time, but one is forced to extend it more and more.

Having begun with biblical concepts appraising the time of the world's existence as several thousand years, science gradually extended the period to at least, a hundred million years since the beginning of organized life and more.[1]

Take the physical history of our solar system alone. We must think of an immense age, of hundred of millions of years, for according to Davy's experiment with the melting of basalt, it would take three hundred and fifty millions of years to cool the earth from 2000° to 200°[2]

Which tells us that even if the history of the world could be confined to such and such number of hundred of millions of years, it would mean only that the universe's era of existence could not be set against or measured with any *real* stretch of time.

As a matter of fact and of necessity, scientific concepts are forced to stretch the above period which, at the first glance, should tend to ultimate limits: foresense uncountable number of years and infinite time.[3]

The antinomy opened by such facts is real. Like's Kant's antinomies, it does arise from the pure reason's ability to show with equal persuasion that certain porpositions can be both proven and refuted.

---

[1] E. Haeckel: *Ueber unsere gegenwärtige Kenntniss vom Ursprung des Menschen*, addendum, § 20.

[2] Helmholtz: *On the Interaction of the Forces of Nature.*

[3] This can be proved also by the fact that neither the insemination of the egg, nor thr formation of premieval plasma could happen at physico-chemical conditions only. Odd as it seems, one has to admit that even the phylogenetic history of animal organisms should have lasted an infinite numbers of years.

In this particular case, it is the experience of our senses that can prove theses which experience does not admit: the infinity of matter and the infinity of the physical universe's history.

The general possibility of such antinomies lie in the situation that we are forced to accept. Being or Existence is not only empirically actual but also transcendentally ideal. Its nature flows from the transcendental forms of knowledge.

XI

We are coming at last to the logician's critique of scientific ideas: to a probing of concepts which are to be the basis for a philosophy distilled from the co-ordinated general scientific views.

Again, and to some extent, one has to consider such ideas formally. One has to specify how much does the clarity and compatibility of such ideas grant them that dual right: to form scientific interpretation of sense-data and also, to present certain sallies into philosophy of science. Further and deeper questions would plunge us into metaphysics.

Our *logical* critique of scientific concepts will have to be but cursory, touching upon one point only: the impossibility of a completely unified and purely scientific view of the universe.

But let us make clear what it should be, that interpretation expressed in a unified scientific view.

Every change or event is conditioned by a set of circumstances which *together* cause the nature of a happening to transpire. The ensemble of such conditions can vary for the same event: be composed of variations or permutations of outer circumstances. Thus, for example, a physical body grows warm when rays of light gathered by a magnifying glass are focussed on it; or when it contracts through blows; or is connected with another body warmer than itself.

But, no matter how variegated the conditions are, the cause of the event becomes manifest with the *simultaneous* presence of these conditions. And that cause is always strictly definite and invariant. In our example, it is a definite quotum of heat strictly related to the degree of heat in the body, to its specific gravity and mass; a quotum of heat equivalent to a definite number of mechanical units.

To define the "*conditions,*" i.e. the group of circumstances attending a change, or more exactly – a new sense-datum, means really only to describe the event. To define the *cause* of a happening, i.e. the quotum of energy which, released or absorbed, seems to "cause" the event, means – to explain it.

The former establishes the scientific fact, the latter – its scientific interpretation. Let us, therefore, turn to the latest and the dominant dynamic concepts.

Causing all changes are the forces "coursing" through matter, their supposed essence being energy. Force is an agent showing hidden or manifest might (potential or kinetic energy). Presenting definite quota of potency, direction of action and point of application, this agent is capable, in one way or another, to "evoke" in a certain body or particle a motion with a certain velocity at a given unit of time.

Let us discuss the first element of action "within" matter, namely the concept of transmission and action of force dimly denoted in the word "cause" ...

This supposed agent does not really exist by itself; it only appears "localized" in a physical body. Therefore, when force sparks a certain velocity in a body or in its particles, it really means only that two bodies find themselves in such a relative position that the dynamic condition of one of them is accompanied by a certain dynamic state in the other. One form of such inter-dependence is action at a distance when there seems to be no intervening physical milieu among the two bodies.

The singular and manifest example of action at a distance is of course, gravitation. All hypotheses "explaining" it – and thus following Newton's famous letter to Bentley – by the difference of pressure in some physical milieu, or by the spreading of some perturbances,[1] all such hypotheses vanish at an even superficial critique. And this, either because of the hypotheses' artificiality forcing upon us a set of suppositions, one less natural than the other, and all less simple than the theory of gravitation; or because of their inner contradiction. Besides, they do explain only a few phenomena.

In addition, the velocity of gravitation, that is, its acceleration, actually unmeasurable as to even its mathematical determinants,[2] points to the force of gravitation's acting at a distance. The magnitude of that velocity does not tend to a definite quotum; rather it signifies that gravitation's velocity can not be even compared with any other.

Which means that, at least in universal gravitation, we perceive a phenomenon totally unexplainable by sheer physico-mechanical conditions; that the physical state of one body is capable of conditioning the state of another object which has no physical connection with the first. Small wonder that physicists tend to explain at least all the rest of examples of visible action at a distance by invisible perturbance in the intermediate physical environment: of ether waves in light and electro-magnetic phenomena, of air waves – in the phenomena of sound.

The majority of actions at a distance is attributed to kinetic or potential energy of particles, that is, to action of intermolecular force or transmission of energy through touch and shock. But the logical critique of the concepts of all such forces can easily show that they act at a distance and differ in only this inessential point: the sphere of their action is smaller. If the very action of molecular force is effected by perturbance in the intermediate physical milieu, then

[1] Clark Maxwell: *Action at a distance.*

[2] According to Laplace's calculations, this velocity is approximately more than six million times greater than the velocity of light.

the latter should be so elastic, continuous, and of such minute weight that its perturbances are spread with tremendous velocity, and in an environment lacking interference – at immense distances.

It is known, however, that molecular forces have a definite radius of action. And if such exists for visible matter, why should not such a radius be present in case of invisible matter, bearer of molecular perturbances? If so, molecular forces manifestly act at a distance: their action is not linked to physico-mathematical moments. Why then have such forces a radius of action?

These considerations in themselves state clearly that molecular forces can not be a purely mechanical principle formed, in the final count, of mathematico-mechanical moments.

The kinetic energy of a body or of its particles, their active force, in other words, is transmitted by shock or touch from body to body and from particle to particle. To such forces are reduced perturbances of the intermediary milieu in some visible actions at a distance.

But analysis of such examples easily shows that the very transmission of an active force is unthinkable without action at a distance. If the live energy of a given particle is a kind of state this particle is in, then one can not see *whose* state it is to be – at the infinitesimal moment of transmission.

At any rate, the very idea of transmission of energy through shock or touch, as a physical principle, does not bear scrutiny.

If the active force of a body is its inner state, it is incomprehensible how that inner state can transmit itself to another body without visible changes. It is equally impossible to see how that inner state of a body can vary not only through physico-mechanical causes but even through mathematical conditions. A resultant velocity, for example, depends on the sinuses of the angles traced by the component velocities.

It is impossible to admit that transmission of energy if but a sparking of force in the other body; for as much quantity of energy is added to the second as is lost in the first.

And so, the critique of concepts concerning action of force "within" matter teaches us that, as a physical principle, they are contradictiory if not senseless. To reduce action at a distance to a transmission of some perturbances in the intermediary milieu means really to sum them up again as action at a distance of small radius. In a physical sense, the latter is as little feasible as the former. The physical state of a body is not comprehensible, unless as a sum of visible or invisible motions of its particles. Even if one admits other physical states not expressible in visible or tactile terms, the connection of such conditions with the mathematical and physico-mechanical moments of events is incomprehensible. [1]

[1] True, one could possibly explain this through concepts of non-euclidean spaces. But is it worth while to explain action at a distance with something, the *actual* meaning of which is still dim, involved and direly relative? There are, however, some willing to take the trouble.

Thus, if concepts of action coming from certain physical or tangible agents make no sense, one finds it necessary to suppose that forces are intangible agents; that their appearance is in no way tied to quota of space, time or any material quantities; that they can be defined by neither quantity, nor direction.

Then, too, the majority of nature laws or concepts of physics lose meaning – such propositions, for example, deemed universal: that the velocity of a body moving through action of some forces changes in connection with the sinuses of angles formed by their directions; or that the force of gravity is proportional to the product of the gravitating masses and inversely proportional to the square of their distance.

Furthermore, if amount or dimensions mean nothing in the case of a non-material agent, then concepts of equivalent forces lose sense, too: such as ideas of the mechanical and chemical equivalent of heat or of termic equivalent of work. Concepts of comparative strain of force lose sense in the same way: the proposition, for instance, that an electric unit attracts billions of times stronger than a one-dimensional unit of gravity [1]; or the idea of decrease of force according to distance.

Finally, it should be easy to show that the ideas of action of forces and quota of energy bring us to a negation of principles that are deduced from these very propositions.

Thus, if matter is infinite – science must accept this even if only because of the implications of the telescope – then so is energy. For each particle of matter is tied to an inextinguishable amount of energy, to forces of linkage and resilience, to begin with. But in this case the second principle of termodynamics, the one concerning the change in entropy, loses validity.

The entropy of the universe tends to the maximum, and, by the same token, the amount of convertible energy – to the minimum. Let no one think of the thesis as a logical trick. It is true, indeed, that if a part of an infinite quantity tends to a maximum, it does not follow that the rest must tend to the minimum. But this proposition does not cover our case of an *actual* quantity, such as energy.

Invalid then becomes also the fourth principle of analytical mechanics, that of the conservation of acting force [2], and with it, the whole building of mechanical physics.[3]

---

[1] Helmholtz: *Contemporary Development of Faraday's Views of Electricity.*

[2] This principle is formulated as $\frac{1}{2}\Sigma mv^2 - P = h$; that is, the differential of the acting energy and the kinetic potential of a system is a constant. Again we think of the world's physical system. Some physicists say that we do not know if this "law" is valid for the sum total of the world's energy. No, in the above case, the logical components of the proposition themselves make no sense.

[3] As it is, we are leaving alone a great number of contradictions between the "laws" of nature and their inter-limitations, the following, for instance. The first principle of termodynamics is expressed in the formula $\frac{P}{Q} = \varepsilon$; $P - \varepsilon Q = 0$ where P presents the quotum of work, Q — of heat, and ε is the mechanical equivalent of heat. This

All of these alogisms derive from the emptiness, if not downward senselessness, of the concept of force or energy accompanied by ideas of quota of force and arrow of its action: in fine, of the action of energy *within* matter, And yet, what seems to admit or even demand such alogical concepts is nature itself.

Here lies the ground for a still another remarkable antinomy which reveals the impossibility of a purely scientific view of the world.

As a sum of physico-mechanical events, the universe seem to bring forth agents the existence of which the universe in its general meaning should not admit. The phenomena of such sum of sense-data should consist of elements of space-time and matter, generally speaking, of elements rational or capable of being rationalized. But forces can not consist of such or be so linked. Force is something entirely irrational.

Science would like to press upon us a contrary opinion, attributing to force some rational traits – action within matter, direction, etc: the very idea that brings about the above alogisms.

This antinomy could be solved, it seems, by the proposition that being an empiric reality, the universe is yet in some way an aspect of transcendental reality; that forces anticipated in the physical nature are but a breath of Being's subliminal substance.

This is why it seems senseless to rationalize force as a purely physical essence by giving it countable determinants: amount, direction, point of application to matter. The concept of force dominating physics – and man's mind – is an odd mixture of some suspected agent, metaphysical and irrational, with physical definitions totally unrelated to the former.

Eliminating from the concept of force those totally inapplicable physico-mathematical determinants, one finds the residue to be but an indivisible, irrational and elemental substance. The doctrine of Empedocles, who like Schopenhauer saw in the forces of nature the strain of will, irrational and metaphysical (even love and hate), this doctrine shows more sagacity than our dyed–in–the–wool scientific concepts.

What does this all really add up to?

First: from science as such it is imperative to subtract all concepts of energy as a physical principle. Purely scientific ideas will then speak to us of matter dead, inert, colorless; of something arrived at by a correct *scientific* epistemology.

Secondly: after such an elimination, the explaining of sense-data formerly accepted in the science of matter, collapses. This science is

proposition, however, is invalid for continuous processes, whose differential formula is $dP - \varepsilon dQ \neq 0$. To help this situation comes the second principle of termodynamics advancing the concept of the non-reversibility of nature's processes and of the loss of energy in heating machines.

then reduced to a mere description of events and of their conditions; loses any autonomous theoretical lining. In this case the preserving in science of dynamic concepts takes on the significance of a mnemotechnical, a working procedure amounting to a mere classification of events.

And the third point. The only elements advanced by dynamic concepts as physical agents, lose sense on closer scrutiny. They can not constitute a basis for empiric cosmology, that is, for a unified, all embracing, yet purely scientific view of the universe. Such a cosmology is even less possible when the science of matter merely describes conditions of events without advancing a clear and non-contradictory interpretation, that is, a harmonized, meticulously co-ordinated system of thoughts.

The above exposé is negative. To exact nature sciences as well as their system of views regarding the world of sense-data, philosophical worth is denied. In scientific gnosticology such value is eliminated by its very viewpoint. A whole and clearly co-ordinate view of the universe is proved to be impossible, as seen in two antinomies whose solution demands: one, of a transcendentally ideal world; the other – of its metaphysical lining.

Only mathematics are thought of as having a certain philosophical value. True, the latter is limited to the more than modest role of a witness testifying to the truth of a philosophical reasoning; and at that, of a witness whom only philosophy can force to speak.

My considerations hide, however, also some positive theses.

Deductions of scientific theory of knowledge, on the one hand, and the necessity for the eliminating from the world of experience energy as a physical agent, point to the urge for a return to the concepts of the Cartesian school.

Furthermore, in energy turning out to be an irrational element, one poorly clad by physico-mathematical concepts, but at the same time a seeming source of events, one should see a breath of the subliminal world.

As a whole, these sequences and their path of reasoning convince us that of *its own strength* science is incapable of a clearly unified and harmonized view of the universe, – even in its narrow limits, even in its conditional and relative viewpoint. Here, physical principles are so throughly enmeshed in metaphysics that only philosophy can disentangle science's doubts.

Only philosophy, the supreme judge of science, can clear the latter of its flagrant shortcomings: selfcontradicting viewpoint; usurpation of mental domaines other than those of science; a limited and narrow general view of Being. To advise science is the duty of philosophy. Science began with philosophy and must end with it.

Index

GPSR Compliance

The European Union's (EU) General Product Safety Regulation (GPSR) is a set of rules that requires consumer products to be safe and our obligations to ensure this.

If you have any concerns about our products, you can contact us on ProductSafety@springernature.com

In case Publisher is established outside the EU, the EU authorized representative is:

Springer Nature Customer Service Center GmbH
Europaplatz 3
69115 Heidelberg, Germany

Batch number: 10830943

Printed by Printforce, the Netherlands